The Association
of American Geographers

The First
Seventy-Five Years
1904–1979

by

Preston E. James and Geoffrey J. Martin

with contributions by
Harm J. de Blij and Clyde F. Kohn

FOREWORD

To celebrate its fiftieth anniversary in 1954 the Association of American Geographers, then recently invigorated by merger with the American Society for Professional Geographers, prepared *American Geography: Inventory and Prospect*. That book constituted a status report on geography as a field of study. Now, after a quarter century of explosive growth, it is entirely appropriate that the status report be on the Association itself. How has the Association discharged its responsibility as a scholarly society? How has it contributed to the advancement of geographical knowledge, and what basis does it provide for further advances in the future?

As Preston E. James and Geoffrey J. Martin show so clearly in this official history, the hopes of the Association's founders have been achieved and surpassed. Creation of the Association of American Geographers was instrumental in the early establishment of the field. Beginning with the focus provided by the Davisian construct, the Association has provided a continuing forum for competitive discussion of ideas, and for the interplay of tradition and innovation. Through the Association's publications, standards of quality have been set and both philosophical and methodological diversity have been fostered.

Today, as throughout its history, the Association continues to seek new ways to support the development of the field. Consistent with what is now a tradition of continuing creative change, the Association is once again, at age seventy-five, reshaping its programs so as to be responsive to changing needs in the years to come.

I trust that all Association members will appreciate this volume as I have done, for in understanding the traditions we have established, we can better appreciate the tools we have created for shaping the future of our chosen field.

Brian J.L. Berry
President,
Association of American
Geographers, 1978-1979

PREFACE

The passage of seventy-five years of Association activity demands a written history. Previous forays into parts of the Association's history have been offered more particularly by Albert Perry Brigham, Charles C. Colby and William Morris Davis. Yet much of the early history of the Association has been lost to the record. None of the forty-eight founding members survive to the present (the last of this group to pass was Robert LeMoyne Barrett, 1871-1969). Scrap books maintained by Albert Perry Brigham and Isaiah Bowman have been lost. Minute books dating from 1904 fortunately have been maintained by a succession of Association secretaries and these have been preserved. Much historical data of value have been gathered and printed in the Association Handbook—Directories of 1956 and 1961. Following establishment of the Association's archival collection in 1968 (now on deposit in the Smithsonian Institution) and the consequent accumulation of further archival material encouraged by the Association's Committee on Archives and History, an official Association history has been made possible.

Intellectual evolution by descent, whether of generation of geographer, office, or committee, exists as a continuum of geographic expression. Our growth has been the product of a seventy-five year intellectual inquisition during which time we sought answers to individual intellectual problems and not to eternal questions. This study of Association history contributes toward an understanding of the history of geography and the functioning of a learned society. Revealed is the contribution which the Association has made to the social and intellectual life of the Nation.

*   *   *

In a study such as this it is hardly possible to acknowledge all the individuals who have contributed both directly and indirectly. We are grateful to all those who have helped.

Especial thanks are due to James R. Glenn, Senior Archivist, of the Smithsonian Institution, who functions as Curator of the Association's ever-growing archival collection. To facilitate this study he has extended service far beyond the cally of duty. Ralph E. Ehrenberg, Association Archivist, has facilitated the growth of this deposit, and has provided help on numerous occasions relative to this undertaking. And to the members of the Association Committee on Archives and Association History who have worked diligently on the subject matter of its title, our thanks are due.

Correspondents who have been helpful as informants include Meredith F. Burrill, Gary S. Dunbar, Herman R. Friis, Otto E. Guthe, Chauncy D. Harris, Richard Hartshorne, E. Cotton Mather, F. Webster McBryde, E. Willard Miller, William D. Pattison, Walter W. Ristow, Glenn T. Trewartha and J. Russel Whitaker.

To the American Geographical Society our thanks for the loan of photographs which are herein reproduced.

Especial thanks are additionally due to two Association stalwarts, Chauncy D. Harris and Walter W. Ristow. Both read the manuscript with care and provided numerous beneficial suggestions for change. Chauncy D. Harris additionally functioned as consultant to the Editor, and Wilma B. Fairchild prepared the index.

Preston E. James, co-author of this book, himself a trove of lore and recollection concerning the history of the Association, has been a source of inspiration in this undertaking.

It is a pleasure to thank Clyde F. Kohn, and Harm J. de Blij each for his chapter, and additional suggestions relating to other parts of the manuscript.

Thanks are also due to Eileen W. James who has drawn the maps showing distribution of the membership in the Association and of graduate departments of geography in the United States, and to my wife, Norma J. Martin, who has functioned as amanuensis to the project.

To J. Warren Nystrom, Salvatore J. Natoli, and Elizabeth Beetschen of the Central Office of the Association, who have administered, helped, and otherwise supported this undertaking, steadfast appreciation.

Geoffrey J. Martin
Editor

Easton, Connecticut
October, 1978.

CONTENTS

ILLUSTRATIONS

1. William Morris Davis, Founder of the Association of American Geographers

I

Prolegomenon: The Emergence of a Field of Learning

In 1935 Charles C. Colby wrote, ''The men who founded our Association were, in their turn, preceded by a long list of scientifically experienced and geographically minded men—a list which carries back to the beginning of our country.''[1] Yet the origins of a considered geography as a field of learning probably belong to ancient Greece. And for centuries thereafter things geographic were noted, and made the object of interest, story-telling, wonder, and from time to time scientific investigation. Philosophers and other thinkers pondered geographic matters. This agar of speculation and knowledge entered literature and thought, some of which was doubtless brought to North America in the early 1600's.

In the New World the inhabitants underwent a wilderness-conquest experience. Hot summers, cold winters, a growing season of length unknown, and an unfamiliar topography were circumstances with which the early settlers were obliged to struggle. Men overcame hardship, or were overcome by it. But this man-fundament struggle led to impression, concern, and thought, about the physical environment. The idea of studying the physical environment was very different from the attempt to subjugate part of the wilderness. An embryonic, nativistic, American geography—perhaps not much more than lore in the first instance—began to evolve resultant to the trek west. Occasionally this thinking was institutionalized in a college level course, though this trickle of learning did not swell until after 1854.[2]

On the international scene the heights of mountains, the depths of oceans, and the bounds of land were being charted. Capes, bays, deeps, fiords and all manner of other phenomena were named after explorers, philanthropists, and royalty. Exploration of the unknown parts of the New World was the order of the day. Colonial empires were being fashioned from lands newly known to Europeans. Study

1

of the world in a manner geographic arose in Europe, and emanated from philosophers, cosmographers, geologists, and others.

Stronger support for the institutionalization of geographic learning came with the creation of geographical societies:[3] Paris, 1821; Berlin, 1827; London (Royal Geographical Society), 1831; Mexico, 1833; Frankfurt, 1836; Brazil, 1838; the Imperial Russian Geographical Society, 1845; and the American Geographical and Statistical Society, 1852. Founded in New York City, the latter was the eighth such society in the world. By 1866 there were 18 genuine geographical societies in existence. Following the Franco-Prussian war of 1870, numerous societies were founded in Europe and by 1900 there were "upwards of 100 such societies in the world, with more than 50,000 members, and over 150 journals were devoted entirely to geographical subjects."[4] Notable among these for the history of American geography were two societies formed in 1888. Both held portent for the development of geography in the United States— The National Geographic Society and the Finnish Geographical Society. The first, with headquarters in Washington, D.C., exerted sway with ever-increasing consequence across the United States, and by providing the apparatus of a professional society began to demonstrate that geography was a serious business. The Finnish Society founded in that same year, 1888, held portent as the first such society to demand demonstrated original research as a prerequisite to membership. This was a feature that captured the imagination and attention of William Morris Davis. Each of these societies began to contribute publications, intellectual leaders, explorers and philosophers.

In 1871 the first of a grand series of international geographical congresses was held in Antwerp. The congress was successful and was repeated at Paris, 1875; Venice, 1881; Paris, 1889; Berne, 1891; London, 1895; Berlin, 1899; Washington, D.C., 1904. Subsequently other congresses were convened, and the International Geographical Union was created. Proceedings of the congresses were published and became a valuable source of geographical information.[5]

Of great significance to the emergence of geography in America were a variety of geographical publications. In the United States the *Bulletin of the American Geographical and Statistical Society* (1852) and the *National Geographic Magazine* (1889) were precursors. From abroad there came, most ·notably, *The Geographical Journal* (1893 but with predecessors dating from 1830), *Erde* (1853), *Petermanns Geographische Mitteilungen* (1855), *Scottish Geographical Magazine* (1885), *Terra* (1888), *Fennia* (1889), *Annales de Géographie* (1891), *Geographische Zeitschrift* (1895), and *Geography* (1901).

Within the United States learning was assuming organization and a structured apparatus. The American Philosophical Society was founded in 1769, the American Academy of Arts and Sciences, 1846, the American Association for the Advancement of Science, 1848. Specializations were evolving. Individuals grappling with the whole of available knowledge began to give way to those who studied in a limited field or fields. National and local learned societies were being established for the humanities, the social sciences, and the physical sciences. By 1908 the *Handbook of Learned Societies* listed approximately 120 national and 550 local societies.[6] Especially significant to the creation of a milieu in which geography could find camaraderie were the Modern Language Association of America (1883), the American Historical Association (1884), the American Economic Association (1885), the American Folklore Society (1888), the American Geological Society (1888), the American Psychological Association (1892), the American Philosophical Association (1901), the American Anthropological Association (1902), the American Political Science Association (1903), the Bibliographical Society of America (1904), the American Sociological Society (1905). The formation of these associations and societies reflected a growing academic interest in the advancement of learning, and created opportunity for the emergence of intellectual consensus by discussion. One society encouraged the emergence of another. The earnest desire for intellectual exchange facilitated association birth and growth. Sometimes a group, sometimes a "school", and sometimes an individual, were responsible for the creation of such an instrument. The growth by membership of these societies was initially confined to the literati, but desire for intellectual accomplishment was across the land. There was in the latter half of the nineteenth century a strong faith in education as a device for increasing equality of opportunity.

University foundation and growth was proceeding apace, and advanced studies more so. In 1861 Yale University offered the first doctoral degree in the United States. The Johns Hopkins University was founded in 1876, Clark University in 1887, and the University of Chicago in 1891, each of which played a vital role in the advancement of graduate work and geography. In the period 1880-1910 graduate education increased considerably and by 1910 some six thousand men and women were enrolled as candidates for advanced degrees. Geography was slow to share in this expansion of graduate study: the subject was perhaps better established in the normal schools. However, in 1897, a department of geography was established at the University of California, Berkeley, and Richard E. Dodge and Miss

Harriet A. Luddington offered sustained geographic instruction at Teachers College, Columbia University.[7] The University of Chicago established a Department of Geography and a doctoral program in 1903. Other universities were soon to add the subject. The formation and founding of the Association of American Geographers (1904) came at a time when geography could receive a helpful impetus in a changing curriculum which began to accommodate the subject.

American Geography: Some nineteenth century antecedents

One of the most pronounced advances of the nineteenth century was an ever-increasing attention to observation by those in the field, and substitution of a genetic treatment for what had once been empiricism. Evolution was embraced. Following the Civil War the West was explored with a newly acquired understanding. Students of the natural order—both organic and inorganic—realized that there was pattern to an ever evolving sequence of forms. Geologists had acquired this viewpoint in the very early 1800's but geographers remained largely empirical for another 50 years.

Numerous intellects contributed to that body of knowledge which geographers read and which they regarded as an intellectual heritage. Lewis Evans, Benjamin Franklin, Lewis and Clark, Dana, and Hitchcock each contributed substantially to "geography" as the observation of the visible facts.

The earliest systematization of the field for American geographers it seems, came with Arnold Guyot's *Earth and Man*. Of this work William Morris Davis later wrote:[8]

> The volume was pervaded by a spirit of rational correlation, and may therefore be regarded as having given us the first great impulse toward the cultivation of geography as a serious and independent science. . . the rational or causal notion was then implanted in our concept of geography . . . the philosophy of the book was. . . teleological.

Fifty years later Nathaniel S. Shaler wrote *Earth and Man in America* (1899), and built upon "the rational correlation of earth and man", introduced in Guyot's book, albeit that Guyot worked under the erroneous principle of biological teleology. The two books undoubtedly were of large import to American geography and geographers of the twentieth century. It was the ontographic and chorologic aspects of geography in North America which were so ignored. The physical environment in contrast received much attention. Lesley of Pennsylvania advanced thought concerning the dependence of form upon structure in numerous works though special attention was accorded his book, *A Manual of Coal and its Topography* (1856). When the United States Geological Survey undertook western surveys, and the Geological Survey of California, under Whitney, investigated its natural resources, numerous topographic maps and survey volumes resulted. Authors, particularly John W. Powell and Grove K. Gilbert, provided generalizations of wide applicability and significance to geographers. And it was the geographic fraternity that provided assistance of critical worth to the geologists. Geographers made the base maps for the King, Hayden, Wheeler, and Powell surveys. "This involved the determination of the geographic coordinates of primary points, the extension of a system of triangulation over the areas in question, and the topographical mapping."[9] John W. Powell wrote, "Sound geological research is based on geography. Without a good topographic map geology cannot even be thoroughly studied, and the publication of the results of geographical investigation is very imperfect without a good map."[10] Blodget provided a useful summary of climate in North America (*Climatology of the United States and of the Temperate Latitudes of the North American Continent*, 1857). Maury advanced our knowledge of oceanic meteorology. Coffin provided *The Winds of the Northern Hemisphere*, 1852. Particularly significant contributions to meteorology were provided by Redfield, Espy, Loomis and Ferrel, and later by the work of the Signal Service of the Army, the Weather Bureau of the Department of Agriculture, the Hydrographic Office of the Navy, and the meteorological observatory at Blue Hill, Boston.

Exploration abroad also added to the fund of geographic knowledge. The U.S. Navy, the Hydrographic Office, journeys of whaling boats, and individual travellers and explorers, all made valuable measurements and observations. Some of these activities contributed toward the development of a variety of human geography. Yet more important for the embryonic anthropo-geography were the studies of Indian tribes by American ethnologists (such as J. Walter Fewkes and W J McGee), the importation of European ideas and thought from university and publication, and studies of commerce and industry by American economists.

Strongly supportive of the geographic undertaking were the studies in geographic geology and physical geography exemplified in Curtis F. Marbut's, *Physical Features of Missouri* (1896), Rollin D Salisbury's, *The Physical Geography of New Jersey* (1898), and Ralph S. Tarr's, *The Physical Geography of New York State* (1902). This was the realm of accomplishment which Davis knew so well, and which he supported so strongly, that led to an emergent scholarly geography.

Geography and Geographers:
The Development of Geography as a Scholarly Discipline

Geography as an academic discipline received a major impetus in Germany in 1874 when the Minister of Education for Prussia directed that each university in that land should establish a chair of geography to be occupied by a scholar with the rank of professor. This decree was soon emulated in other parts of Germany. Then it spread to other countries, especially on the continent of Europe.[11] Academic geography began to develop in the universities of the United States somewhat later.

In Germany in 1871 geography was represented by chairs at only three universities (at Berlin by Heinrich Kiepert, at Göttingen by Johann E. Wappäus, and at Breslau by Karl J.H. Neumann). But at that time none of the universities offered students a program of advanced training in geography as a discipline. Skilled map-makers trained in map-publishing houses, and scholars educated originally in other fields, turned their attention to problems concerning the earth as the home of man. There already existed a long record of interest in studying and writing geography but the scholars of this period were not trained as specialists in geographical studies. For them geography was only one facet of scholarly careers aimed at the advancement of learning. The appointment of distinguished scholars to professorships of geography received the attention of scholars everywhere: Oscar Peschel to Leipzig (1871), Alfred Kirchoff to Halle (1873), Hermann Wagner to Konigsberg (1875), Georg K.C. Gerland to Strasburg (1875), Friedrich Ratzel to Munich (1875), and Ferdinand von Richthofen to Bonn (1877). The geographers appointed in Germany after 1874 brought with them a variety of viewpoints: many had been geologists; some had been botanists or zoologists; some had been journalists, or editors. Of the 21 universities of the German Empire in 1891, 19 offered geography; there were then 45 professors and privatdocenten (in 1885 six years earlier that number was only 26).

Each newly appointed professor had to seek his own answer to the question—what is geography? From 1880 to 1914 the German periodicals presented many different answers to this question.[12] Out of the resulting discussion a point of view was established for geographical scholarship. These standards were transmitted to other countries where they were modified, perhaps added to native contributions or perhaps rejected. As a result each national ''school'' embodied something from the German corpus of thought.

The ''new geography'' that began to appear in the United States in the 1880's was the product of ideas from a variety of sources. The underlying structure came in part from Germany. German ideas were brought to America by European scholars who had studied in Germany and then came to the United States: Arnold H. Guyot, for example introduced ideas from Carl Ritter when he lectured in Boston in 1848, and when he taught at the College of New Jersey (later Princeton) from 1854 to 1884. German ideas were also brought back to the United States by Americans who went to Germany for all or part of their advanced training—scholars such as Daniel Coit Gilman, Ellen Churchill Semple, Emory R. Johnson, J. Russell Smith, Rollin D Salisbury, and Charles T. McFarlane. The French ''possibilists,'' especially Vidal de la Blache and Jean Brunhes, were widely read by Americans, as were the British geographers, especially Herbert J. Fleure, Andrew J. Herbertson and Halford J. Mackinder.

But there were also certain distinctive and indigenous American ideas that were built into the structure of the emerging professional field which were derived from such American scholars as John Wesley Powell, Clarence King, and Grove Karl Gilbert. In the United States geography seems to have reached a stage of emergence into a self-sustained growth in the middle 1890's. John K. Wright adopts 1895 as the divide in American geography between lore and a growing science.[13] From an encyclopaedic approach of the earlier 1800's geography evolved in the second half of the nineteenth century into a larger subject groping its way via teleology and evolution, concerning itself with natural law, and adopting a loosely knit form of environmental determinism as an organizational framework for the subject. The nineties witnessed a stage of geologic investigation in which man was perceived as being part of the last phase of earth history. In studying the recent geologic past an embryonic human geography began to emerge in the mental construct. Some geologists began to make studies that were geographic in character.

American Geography: A Discipline Emerges

It has been suggested[14] that before a field of learning can become a learned profession four supporting circumstances are necessary: (1) a number of scholars actively working on related problems, and in close enough contact with each other that ideas are quickly disseminated, and critical discussion stimulated; (2) departments in universities offering advanced instruction in the concepts and methods of the field; (3) opportunities for qualified scholars to find paid employment in work related to the profession; and (4) an organization, such as a professional society, to serve the interests of the profession and provide a focus for professional activities. In the United States the Association of American Geographers provided this fourth support. It was necessary also to establish the objectives and methods of the field of geography. "Each active scientific discipline," writes Wilbur Zelinsky, "must arrive at its own peculiar definition of acceptable problems and acceptable ways of attacking them through a social process of mutual consent."[15] This is what has been called the process of competitive discussion. The way in which a consensus is to be reached cannot be set forth in by-laws; it cannot be decided by a majority vote, nor by a decree from someone in authority. Mutual consent suggests a loose consensus, continually subject to revision. This common paradigm seems to be a necessary pre-requisite to research which takes its place in an intellectual structure. Prior to the creation of the paradigm, or the advancement of a corpus of thought to paradigm candidacy, fact-gathering and thought compilation is random and frequently localized. A natural history of geography emerged in the nineteenth century in the absence of a meshing of theoretical and methodological belief which would have permitted selection, evaluation and criticism . . . in short, scientific development. There was no external metaphysic, no one individual or historic incident which functioned as a thesis which would begin competitive discussion among qualified scholars.

Yet toward the end of the nineteenth century, such an individual— William Morris Davis—emerged with a theoretical construct. He offered a thesis which could be discussed, revised, opposed, but hardly ignored. For a quarter of a century geographers in the United States largely agreed with the Davisian schema. Transition to maturity occurred. In no small measure that was due to Davis's commanding comprehension of the field. Head and shoulders above his contemporaries, he was authoritative and authoritarian.[16] Although his thought was challenged by Walther Penck[17] he seemed not to have been substantially and successfully challenged in the United States until Sauer posed *The Morphology of Landscape* (1925), in opposition to the scheme of Davisian causation.[18] In consequence American geography and geographers were offered the prospect of disciplinal growth devoid of major ideological schism or personal faction. Disagreements there were, but some 25 years of reflection, discussion, and refinement were experienced before the first paradigm was tumbled and quiet scientific revolution ensued. Often such a challenge to the paradigm comes from the younger scholars, and especially from those few who possess unusual capacity . . . perhaps imagination and eloquence. (Sauer was thirty-six years of age in 1925).[19] The older scholars seldom change their minds, but the usual experience is that the objectives they set for themselves prove unreachable. The matrix of society changes, technology advances, knowledge is increased, and demands for knowledge application change. Marvin W. Mikesell has written "If one looks back, the landscape of American geography may seem littered with the detritus of abandoned tasks."[20] The older scholars may be frustrated and disgruntled when they see the tasks they started left unfinished, perhaps not even tested. But the process by which new goals and new procedures are suggested, is stimulated by the give and take of competitive discussion. The greater the number of participants in discussions the more rapidly new parameters may be defined, and the more frequently does the profession seem to change its course. To identify one single model of professional work becomes impossible. Such a process may seem loose and inconclusive—as, indeed, it is. But for a learned profession in process of growth it is a highly effective procedure.

The formation of the Association of American Geographers in 1904 facilitated scholarly discussion which led to assent and adumbration of Davisian physiography (and concomitant ontography), and new viewpoints of the twenties . . . which was American geography in the making.

2
American Geography in Universities and Geographical Societies

Two conditions which encourage an on-going competitive discussion of concepts and methods are graduate training programs in universities, and employment opportunities for qualified scholars. In the United States three universities were of major importance in starting such programs. Although they did not award advanced degrees in geography, they were based on studies which later became known by that name. Harvard, with William Morris Davis and Nathaniel Southgate Shaler, was the first to offer special training in physical geography. Yale University appointed Daniel Coit Gilman as Professor of Physical and Political Geography, but as he soon left to become President of the University of California (and later of the Johns Hopkins University), he was succeeded by William Henry Brewer (1873), Francis A. Walker, and in 1898 by Herbert E. Gregory. The Wharton School of Finance and Commerce, at the University of Pennsylvania, began to offer advanced training in economic geography in the 1890's, initially with Emory R. Johnson and then with J. Russell Smith as instructors. Graduate work leading to higher degrees was also offered at many other universities including most notably the University of Chicago. But a geographic offering, quite additional to courses in physiography, had already begun to emerge at a number of universities and colleges regardless of departmental title.

Departments (of geology, geology and geography, or geography) which offered six or more geography courses included:[1] Columbia University (1899), Cornell University (1902), University of California (1903), University of Chicago (1903), University of Nebraska (1905), Miami University, Ohio (1906), University of Minnesota (1910), University of Pittsburgh (1910), Nebraska Wesleyan University (1911), University of Wisconsin (1911), Harvard University (1911),

University of Pennsylvania (1913), New York University (1913), Yale University (1914), Denison University (1914), University of Michigan (1917), Ohio State University (1918), University of Missouri (1918), Northwestern University (1919), and the University of Illinois (1920).

Of these universities, Harvard, Yale, Pennsylvania and Chicago, were perhaps the major sources of the ideas that were involved in the competitive discussion in the first decade after 1904.

William Morris Davis at Harvard

Harvard was the first American university to offer a program of advanced training for future geographers. Since the prime mover in the development of geography as an academic field in the United States was the Harvard professor, William Morris Davis, we need to know more about him and the sources of his ideas.[2]

When Davis entered Harvard as an undergraduate it seems he had no clear idea of what he wanted to do with his life. Three of his teachers Raphael Pumpelly, Josiah Dwight Whitney and Nathaniel Southgate Shaler "may have influenced his final choice of a career."[3] Although Davis studied under and later worked with Pumpelly, it was the ex-Kentucky cavalry man and poet-scientist Shaler who seems to have most fascinated him. After graduation with the class of 1869 Davis took a year of graduate study leading to a Master's degree in Engineering. He then accepted a position as an assistant to Benjamin A. Gould (Harvard 1844), who had been appointed director of the new observatory at Cordoba, Argentina. There Davis remained for three years before returning to his native Philadelphia, where he took a position as a bookkeeper in the Barclay Coal Company (1873-1876). Missing the intellectual stimulation that his restless mind demanded, Davis enrolled in a field course in physical geography which Shaler was offering at a summer camp, Cumberland Gap, Kentucky. Davis distinguished himself in the course, whereupon he was invited to return to Cambridge as Shaler's assistant in field geology.

Following a year of teaching and studying with Shaler, Davis embarked on an around-the-world trip with his cousin, Thomas Mott Osborne. The two began their travels in September 1877, journeying across the western part of the United States, thence to Japan, China, India, Egypt, and western Europe. Davis returned to Cambridge in September 1878, and was appointed Instructor in Physical Geography.

A very interesting sequence of events followed. After four years of teaching with Shaler, the time came for his reappointment. In June

1882, Davis received a letter from President Charles W. Eliot, on behalf of the Harvard Corporation, agreeing to reappoint him, but warning "this position is not suitable for you as a permanency. . .the chances of advancement for you are by no means good. . ." Davis himself felt that his teaching lacked the excitement that comes from the search for explanations and the use of general concepts. That summer Raphael Pumpelly, a member of the geology staff at Harvard, offered Davis a job on a field survey which he was supervising in Montana. The purpose was to identify and map the resources that might provide revenue for the Northern Pacific Railroad. Davis was assigned the task of studying the geological formations surrounding the coal measures of Montana. It was in the course of this work that Davis recognized an erosion surface of slight relief, preserved in terraces on either side of the Missouri River. He was familiar with John Wesley Powell's work describing such surfaces, and relating them to the base level of drainage. If the Missouri River had cut its valley below this surface there must have been a general uplift, thus permitting the streams to deepen their valleys. Later Davis said that it was in Montana that the outline of the cycle of erosion first became clear to him. With this theoretical construct functioning as an organizational device for further physiographic study, Davis's research and teaching assumed new meaning and vitality. His teaching at Harvard, when he returned in the fall, was greatly improved.[4]

Davis became well acquainted with the leading geologists of that time. He visited Washington, D.C., where Powell, Director of the United States Geological Survey, offered him a job. But Davis preferred to return to Harvard to "fight it out" for an improved status there, though he worked with Thomas C. Chamberlin during summers for the Survey, on a study of New England.

In 1885, just three years after President Eliot had warned him that he had a restricted future at Harvard, he was appointed Assistant Professor of Physical Geography. It was at this time that Shaler and Davis began to offer summer field courses for advanced students.[5] The Department of Geology was soon strengthened with the addition of Reginald A. Daly (1893), Jay B. Woodworth (1893), Robert DeCourcy Ward (1895), and Robert T. Jackson (1899). This provided Davis with the opportunity to offer "Physical Geography of the United States," and "Physical Geography of Europe" in addition to other systematic courses in physical geography. Frequently, in his courses, Davis lectured from physical cause to human consequence.

At what stage did Davis become interested in the formation of a scholarly field called geography? It is unlikely that there was any one moment when Davis decided to assume leadership for this purpose, or any one source from which he derived the idea. Perhaps the most influential source was his teacher, Professor Shaler. Shaler was primarily concerned with the earth as the home of man, and frequently included ideas of such relationships in his lectures and in his writings. In this way he directed the thoughts of his students and colleagues toward the many interconnections between the physical earth and its human inhabitants. Yet it is curious that Davis seems not to have read Powell's studies of Indian cultures, and especially his program for changing the patterns of settlement in the lands of the arid region.[6] No more striking example of a study of the relation of human settlement to the physical character of the land could have been provided. He knew Powell personally, and adopted his ideas concerning the classification of rivers and the process of denudation by running water to a regional base level. It is also surprising perhaps that Davis makes no reference to Ferdinand von Richthofen, whose inaugural lecture at Leipzig was published in 1883, and includes just the kind of statement on the nature of geography that he was seeking. Even as late as 1889, Davis described geography in terms which offered no clear statement concerning human geography. In an article published in the first volume of the *National Geographic Magazine,* (1889), he writes that geography should[7]

". . .include not only a description and statistical account of the present surface of the earth, but also a systematic classification of the features of the earth's surface, viewed as a result of certain processes acting for various periods, at different ages, on diverse structures."

Davis was much impressed with what Arnold Guyot had to say about the teaching of geography. The newest idea in the field of education in the late eighteenth century was that memorizing words and statistical data could have little lasting value. Ritter insisted that the "new scientific geography" must seek to identify the causes of observed features. As Ritter's disciple, Arnold Guyot, put it, the earth should be presented to students as a "complex system of harmoniously related parts." He wrote that[8]

"The great geographical constituents of our planet—the solid land, the oceans, the atmosphere—are mutually dependent and connected by incessant action and reaction upon one another; and hence the earth is really a wonderful mechanism all parts of which work together harmoniously to accomplish the purpose assigned to it by an all-wise Creator."

Davis followed the teaching of Shaler and Guyot, and believed with them that geography should seek explanations and attempt the prediction of consequences. Both Shaler and Davis rejected Guyot's teleological explanations. The brilliant studies of Charles Darwin and Alfred Russel Wallace had introduced to the world of science the ideas of evolution. Davis extended the concept of evolution by analogy to changes in the physical features of the earth's surface. His concept of the cycle of erosion was what we would today call a "theoretical model." Davis continued to work diligently to extend and improve his model, and also to provide a simple description of it suitable for teaching in the schools. In 1889 he gave a lecture before the Science Association of the Johns Hopkins University in which he outlined a relatively simple model of landform development that could replace the bewildering description of details commonly presented in secondary schools.[9]

Davis himself was primarily a physical geographer, or dynamic geologist. But the more he tried to extend his ideas to secondary school education, the more he realized that the treatment of the physical earth, developed by physical processes, was insufficient. He sought a meaningful way to relate studies of the physical environment to all forms of life including the human inhabitants. This part of his scheme Davis later described as ontography, to distinguish it from the physical features and processes which he called *physiography* (a term which he had borrowed from Thomas H. Huxley).[10] His solution in the early 1890s was to apply Darwin's biological concepts by analogy to human societies. From the English philosopher, Herbert Spencer, he began to picture human societies as comparable to organisms, struggling for survival by adjustment to the physical environment. This form of Social Darwinism seemed to offer a way to bring people trained in the physical sciences together with people trained in the social sciences.

In 1892, the National Education Association appointed a committee to consider the related problems of the content of secondary school education and college entrance requirements. The Committee of Ten, as it was called, was chaired by Charles W. Eliot, President of Harvard.[11] The Committee appointed nine conferences, each to review the content of secondary school courses of study in a recognized field, and to relate course content to entrance requirements for colleges. The Conference on Geography was headed by Thomas C. Chamberlin who, in that same year, resigned as President of the University of Wisconsin and assumed Chairmanship of the Department of Geology at the newly formed University of Chicago. There were eight members of the Geography Conference, including Davis and others selected from universities, teacher-training schools, and secondary schools. In retrospect it appears that Davis was trying to formulate the idea of an earth science organized in terms of a dynamic model of earth-forming processes to which the patterns of human settlement might be related. At this time he had not, however, constructed a point of view that might bring physical geographers and human geographers together into a coherent group. Ontography at this time had not been well developed, had not been made the subject of publication, and was therefore not a unifying theme. Undoubtedly Davis learned much from the discussions at this conference, and he made effective use of his skill in putting together ideas from diverse sources into a simple and unified plan. Israel C. Russell (of the Department of Geology at the University of Michigan), a member of the conference, believed that Davis's ideas dominated the resulting report which was later adopted by the Committee of Ten.

Davis also prepared text books and teachers guides that presented the new content in a form suitable for use in the schools. Only a small proportion of the schools in the United States followed the recommendations, and in less than ten years the program that Davis had formulated was recognized as a failure. Davis could not reach enough teachers to impart his knowledge and his enthusiasm, and most teachers remained unprepared to teach anything novel in the field of geography. The ideas of Social Darwinism, however, persisted in the schools long after the courses of study Davis had devised were abandoned.

Nevertheless, Davis's incursion into the field of secondary education may have been a decisive factor in leading him to attempt definition of a new professional field embracing the physical aspects of the earth as well as the obviously related elements of human geography. In 1892 Davis could not have persuaded secondary schools to accept a program of pure physical geography, but with the activities of man added, the program appealed to many who were resisting the trend toward increased specialization. The formula that appealed to a few school teachers also made it possible for specialists in physical geography to find some kind of common ground with those who pronounced geography a social science. Lacking in 1900 was a professional society where ideas such as those formulated by Davis might have been discussed among qualified people.

During his years at Harvard, Davis trained many young scholars who later joined him in his efforts to organize a professional society for geographers. His students found employment in many universities, often to teach courses in physical geography in departments of geology. His students also became significant figures in several agencies of the federal government. His student record is remarkable in the history of American geography.

Geography at Yale

Jedidiah Morse ("the Father of American Geography") had brought attention to geography at Yale, through his election to a tutorial post in 1786, and subsequent adoption of his texts at the university. Although geography was dropped from the Yale curriculum in 1825, the subject was a university entrance requirement from 1826 to 1881. In 1863 Daniel Coit Gilman, at the age of 32, was appointed Professor of Physical and Political Geography at Yale. Gilman's contribution was largely physiographic, though he was concerned to place the physical environment in the world setting.[12] When Gilman departed Yale to assume the presidency of the University of California in 1872, William H. Brewer replaced him. Brewer was among the first of U.S. geographers "to consider the topography of the earth as an influence on the tribal movements and habits of man." Also in 1872, Francis A. Walker (superintendent of the U.S. Census Office, under whom the ninth census of the United States was conducted in 1870), was appointed to the Yale faculty. Walker was fundamentally concerned with the distribution and movement of population, and the "center of population" concept.[13] When he left Yale in 1881 to assume the presidency of the Massachusetts Institute of Technology, geography had established for itself a position on the campus. The geologists were not unfriendly to the geographical point of view. Early forms of determinism were the subject of lectures in anthropology by William Graham Sumner. In 1898 Herbert E. Gregory was appointed to the Yale faculty. He had been graduated from the College in 1896 and was familiar with the geographic offering. Consequently he built upon what already existed and did not attempt a radical departure. He swiftly built a department of geography "within the geology department", that nurtured Davis's ontographic idea. From it there emerged, in the decade 1900-1910, a very strong regional offering.[14] The Yale contribution, substantially ontographic in point of view, supplemented the physiography at Harvard.

Emory R. Johnson at Pennsylvania

Meanwhile a very different approach to the formation of a geographic profession was emerging at the University of Pennsylvania's Wharton School of Finance and Commerce. Here leadership in the development of geography was provided by the economist, Emory R. Johnson. The approach pioneered by Johnson was antithetical to the Davisian schema. It inverted the Davis formula to give primary emphasis to man's production and transportation of goods in relation to the patterns of physical features and resources on the earth's surface. There were, Johnson believed, certain economic principles that could be used to describe the relationship of man to the land. In this formula "man" came first.

The unseen figure in this approach to geography through economics rather than geology was the economist Richard T. Ely who taught political economy at Johns Hopkins from 1881 to 1892, and at the University of Wisconsin from 1892 to 1925. He spent four years in Germany, from 1876 to 1880, mostly at Heidelberg, studying with a leading "historical economist." This was the period when geography was developing in Germany, and geographical ideas were being widely transmitted to related fields of study. Although no geographer was appointed at Heidelberg until after Ely had completed his work there, many economists at that time were already convinced of the importance of geography to their own studies. When Ely returned to the United States he was appointed to the faculty of the Johns Hopkins University. Among his students was Emory Johnson, who derived from Ely many ideas about the importance of a sound foundation of geographical study. Ely also taught Henry C. Taylor, L.C. Gray, and M.L. Wilson, all of whom contributed much to the development of land economics during the present century. At Wisconsin one of Ely's outstanding students was Oliver E. Baker.

Emory Johnson studied with Ely in 1890-1891. Then after a year in Germany, including several trips to other European countries to observe the methods of operating and maintaining inland waterways, Johnson returned to the Wharton School where he was awarded a Ph.D. degree in 1893. His dissertation—"Inland Waterways: Their Relation to Transportation"—was identified by Whittlesey in 1935 as the first in the history of American geography to concern a geographical problem.[15] In that same year (1893) Johnson was appointed to the faculty of the Wharton School,[16] and, in 1894, when a four-year curriculum was adopted at Wharton, Johnson introduced required undergraduate courses in physical, economic, and commercial geography. In 1896 he offered his first post-graduate course, "Theory of Transportation." By 1897 Johnson offered "Theory and Geography of Commerce" (the text was G.G. Chisholm *Commercial Geography*), "Physical and Economic Geography" (texts were Tarr's *Elementary Physical Geography* and Monographs 1—X of the National Geographic Society), "Transportation," and "Commerce." By 1901 Johnson had added "American Commerce and Commercial Relations" and "European Commerce and Commercial Relations", and J. Paul Goode was teaching "Economic Geography of America." The following year Goode offered "Economics," "Geomorphy, with Economic Applications," "Climatology, with applications in Economic Geography," "Political Geography," "Economic Geography of America", and "Economic Geography of Europe."

Johnson's reputation as a transportation economist spread rapidly. In 1899, after the Spanish-American War had demonstrated the need for an inter-ocean canal across the Isthmus of Panama, the United States Congress appointed an Isthmian Canal Commission to provide cost-benefit analyses of alternative routes. Johnson, a member of the Commission, was assigned the task of working out a forecast of the amount of traffic that would use the canal in 1914—the year when it was to be completed. Johnson selected as his assistant an advanced graduate student, J. Russell Smith.[17] Johnson and Smith had had no previous training in the geography of commerce: they found it necessary to work out their own methods and to gather their own data. Johnson's forecast, estimating the traffic that could be carried in 1915, proved to be remarkably accurate. This experience convinced Johnson and Smith that geography was an essential foundation for the analysis of economic problems. In 1901, with the Commission work completed, Johnson returned to the Wharton School, and Smith left for a year of study with Ratzel in Germany.

During the forty years Johnson was at the Wharton School (1893-1933, 1919-1933 as Dean), he trained many of America's leading economic geographers. The first to complete a dissertation under his direction was J. Paul Goode, in 1901. Goode had studied with Davis in 1894, and had been a Fellow in Geology with Thomas C. Chamberlin and Rollin D Salisbury at Chicago in 1896-1897. When he found that he could not secure an advanced degree in geography at Chicago he went to the Wharton School to earn the degree in economics. At this time Goode may have been the only student of geography in the United States who had taken work with William Morris Davis at Harvard, Emory R. Johnson at Pennsylvania, and Rollin D Salisbury at Chicago. In 1901 Johnson established a subdivision of his Department of Economics and Social Science, which was described as "Geography and Commerce." Goode was named chairman.

In 1903 Goode left the Wharton School to join Rollin D Salisbury in the new Department of Geography at the University of Chicago. J.

Russell Smith had by this time returned from Germany, and had completed his dissertation at Wharton on "The Organization of Ocean Commerce." Johnson appointed Smith chairman of what became known as the Department of Geography and Industry, which had by then been separated from the Department of Economics and Social Science. The Department of Geography and Industry can properly be termed Smith's creation. Three years later Walter S. Tower joined Smith, and, by 1907, courses on "Earth and Man," "Climate and Civilization," "Economic Conditions of South America," "Field Work in Industry," and "Industrial Management" had been added.

During this period there was very little contact between geographers working at Harvard with a foundation in geology, geographers at Yale with the ontographic viewpoint, and geographers working at Pennsylvania with a foundation in economics. Travel between these places was not lightly undertaken, and certainly not for casual purposes. Even if geographers from these three institutions had been brought together physically they would have found some difficulty in exchanging ideas. Until a way could be found to overcome these obstacles, no real competitive discussion of what might constitute the conceptual structure, and the acceptable procedures for defining problems and seeking answers, could be started. The lack of contact between the three groups was clearly revealed by Davis, some twenty years later, in his paper on "The Progress of Geography in the United States:"[18]

> "The chief impulse toward the higher study of economic and commercial aspects of geography in America appears to have come, about 1900, from the Wharton School of Finance and Commerce, in the University of Pennsylvania."

Other New Departments of Geography

Note should be made of the numerous other universities and colleges where geographical work within a department of geology was undertaken or where departments of geography were established before 1904. At the Johns Hopkins University in Baltimore, where Daniel Coit Gilman was president, and where academic standards that had been pioneered in Germany were introduced in the United States in 1876, graduate work leading to the Ph.D. degree in physical geography, including meteorology and climatology, was possible. George B. Shattuck, who was described as a physical geographer and explorer, Harry Fielding Reid, a seismologist who surveyed Alaskan glaciers, and Oliver L. Fassig, meteorologist and climatologist, were the "geographers" of the staff in 1903.

In 1898 George Davidson was appointed Professor of Geography in a new Department of Geography at the University of California in Berkeley. The work at Berkeley, like that at Johns Hopkins, was all in various fields of physical geography, including meteorology. In both universities there were close connections with geology. Berkeley awarded no Ph.D.'s during this period, but the University of California had the first Department of Geography (not the binomial Geology and Geography) in any major American university. In 1897 Teachers College of Columbia University offered four geography courses. By 1903 Cornell offered a dozen courses in geography within a department of geology. Encouragingly growth in the number of geography courses offered, and in the number of students taking these geography courses, increased. By 1910 twenty-four of the then forty state universities offered courses in geography. In that year the University of Chicago enrolled more than 800 geography students annually. The Universities of California and Wisconsin could count more than 500 each, Yale had 400, Pennsylvania 260, and 12 other universities counted not fewer than 100 students each.[19] There were also many departments of geography in normal schools and teachers' colleges, but none of these provided post-graduate training in geographic research.

The Year 1903

The preliminary steps toward the formation of a scholarly geographic organization in the United States reached a climax in 1903. The number and variety of published contributions, each the product of geographic research, increased rapidly. Two important books, published in 1903, dealt with the influence of the earth's physical features on the course of American history. The authors were Albert Perry Brigham and Ellen Churchill Semple. The first department of geography offering a graduate program in geography leading to the Ph.D. degree was established at Chicago. But there was still no scholarly society where mature scholars, who had made original contributions to the study of geography, could meet for the kind of competitive discussion that could lead to consensus regarding the objectives and limits of this field of study.

The address by Davis, in December 1903, proposing specific steps for organizing a professional society, and the procedure followed, during 1904 leading to the first meeting in Philadelphia in December 1904, will be examined in the next chapter.

The Increase of Studies in Economic and Political Geography

Students trained by Davis had started to produce scholarly papers in physical geography some years before works by Johnson's students appeared. To be sure, Johnson had, in 1893, written the first doctoral dissertation on a geographic problem. His students, J. Paul Goode and J. Russell Smith, published papers in economic geography in 1901. J. Paul Goode's doctoral dissertation was entitled "The Influence of Physiographic Factors Upon the Occupations and Economic Development in the United States,"[20] and J. Russell Smith's publication was entitled "Western South America and Its Relation to American Trade."[21] The first paper on economic geography of an independent country (Argentina) was published by Smith in 1903,[22] and, in January 1904, he contributed a similar study on Chile.[23] In 1905 Smith's dissertation, "The Organization of Ocean Commerce," was published.[24] In 1903, Israel C. Russell published the first American paper on political geography.[25]

Studies in "Geographic Influence"

Albert Perry Brigham, one of Davis's earliest graduate students, was appointed to the staff of the Department of Geology at Colgate University, after having completed work for the M.S. degree at Harvard. He taught geography at Colgate for 33 years, and became one of Davis's most devoted helpers. Later, Brigham became a keen critic of the treatment of "geographic influences" by geographers. In 1903 he published a book entitled *Geographic Influences in American History*. Brigham's treatment of the subject was careful, thoughtful, and well written. His treatment of the origin of United States landforms was well received but his meager background in history was reflected in a less thorough treatment of this part of his subject, as historians were quick to note. Brigham, who was a master of English prose, produced a very readable account of the geologic origin of surface features. A most interesting review of Brigham's book by Emory Johnson compared his treatment with that of Ellen Churchill Semple, who wrote on the same subject.

Ellen C. Semple was born and brought up in Louisville, Kentucky. She graduated from Vassar in 1882, at the age of nineteen, having majored in history, political economy, classical languages, and

English. While teaching ancient history in a girls school, in Louisville, she undertook a program of reading in sociology and economics, and completed a thesis "Slavery: A Study in Sociology" which earned her the master of arts degree at Vassar, in 1891. She had meanwhile heard of a professor in Leipzig, Germany, who was offering an exciting new interpretation of history. She, accordingly, went to Leipzig where she studied in 1891-1892, and again in 1895, with Friedrich Ratzel, professor of anthropogeography. She decided to devote herself to the task of making Ratzel's ideas known in English-speaking countries.[26] When she returned to the United States, Semple began to study and write about the ways the physical setting of America influenced the course of American history. What became one of her most celebrated papers was an "explanatory description" of the human communities of the Cumberland Plateau of eastern Kentucky. In 1901 Semple published "Anglo-Saxons of the Kentucky Mountains."[27] This paper constituted an innovation in American geography because it was based on the direct observation of human communities and the results of their isolation.

In 1903 she completed a manuscript concerning the influence of the physical features of the earth on the course of American history. She adopted the title *Geographic Influences in American History* but subsequently changed it to *American History and Its Geographic Conditions* upon learning of Brigham's manuscript. Neither Brigham nor Semple was aware of the other's plan, and there is even some indication that they were about to publish books with the same title when the two Boston publishers discovered the coincidence.[28] Although Brigham and Semple wrote on a similar subject, and from much the same point of view, the emphasis each gave to the various aspects of the problem was different. As noted, reviews of Brigham's book stressed the inadequate treatment of history. In a review of Semple's book, Emory R. Johnson commented that:[29]

> . . .the author evidences a good knowledge of American history. Her information regarding transportation and industry is less thorough.

Numerous other geographers began to write in terms of "influence." Most notable among these, perhaps, were Ray H. Whitbeck, Richard E. Dodge, and the young Ellsworth Huntington. Although Semple had not studied with Davis, Gregory, or Johnson she was already known through her publications. By 1903, when *American History and Its Geographic Conditions* appeared, she had published several articles on geographic matters, most of these in the

Journal of Geography. In 1903, the editor of the *Journal* was Richard E. Dodge, and Davis was a member of the editorial board. Ellen Semple had already made an impression on these leaders, even before any of them had met her.

The study of influences, controls, adjustments, and determinisms was the inevitable result of adding ontography to physiography. In the United States this point of view went through a period of acceptance, doubt, and rejection by a majority opinion of geographers. The paradigm had shifted, and growth had taken place. The subject of "influence" had been instrumental in this.

Geography at the University of Chicago

The University of Chicago was an entirely new creation when it was founded in 1891. The first President, William Rainey Harper, undertook to build a university similar to Johns Hopkins with a faculty of outstanding scholars. He recruited such distinguished individuals as Thorstein Veblen in economics, Albert Michelson in physics, Thomas C. Chamberlin in geology, and Albion W. Small in sociology. Chamberlin resigned the Presidency of the University of Wisconsin to join the faculty at Chicago, and he brought with him from Wisconsin a brilliant young geologist, Rollin D Salisbury. The search for academic talent continued. In 1902 the distinguished British geographer, Halford J. Mackinder visited Chicago, and President Harper offered him an appointment in geography. Mackinder declined. Initially, physical geography was offered in the geology department. Courses closely related to geography were offered in the departments of zoology, botany, political economy, and history. In 1903 an independent Department of Geography was established, with Rollin D Salisbury, Dean of the Ogden Graduate School of Science, as chairman, with an emphasis on graduate work.[30] Salisbury retained close connections with geology, and the two departments shared the same building. In the initial announcement it was stated that the Department of Geography was not to offer courses that were already given in other departments, but was to develop new offerings intermediate between geology and climatology (in the physical sciences) on the one hand, and history, sociology, political economy, and biology (in the social and biological sciences) on the other.[31]

At this time Salisbury remembered J. Paul Goode, who had been a graduate student at Chicago in 1896-1897, and who had greatly impressed Salisbury with sound ideas about the potential field of geography. Goode, who had gone to Pennsylvania to earn the Ph.D.

degree with Johnson, was invited by Salisbury to draw up a plan for undergraduate and graduate programs, listing courses that might be included in geography, and accompanying the list with a memorandum justifying the selection. Goode did this in considerable detail, whereupon Salisbury offered him an appointment at Chicago. Goode joined the geography staff under Salisbury for the second term of the summer session, in July 1903.

What was the program that Goode recommended for the new department?[32] The original manuscript, written on July 17, 1902, and submitted by Goode to Salisbury, included courses in physiography, mathematical geography, phytogeography, zoogeography, paleogeography, anthropogeography, economic geography, history of geography, political geography, regional geography, geography teaching, and field courses. When the Department of Geography opened its doors in July 1903, its courses were essentially those proposed by Goode. The accomplishment was as significant as it was considerable, establishing geography nearly a thousand miles west of the Harvard, Yale, and Pennsylvania accomplishments, and showing that, at Chicago, intellectuals felt that to be a geographer was a serious business.

A few years later, the sociologist, E.C. Hayes, wrote to J. Paul Goode for his ideas concerning the content and objectives of what was then called the "new geography." The essay that Hayes wrote was published in 1908, and we may assume, therefore, that Goode wrote the following reply to Hayes's question in 1906 or 1907:[33]

"The matter of largest interest in modern geography is the interaction between man and his physical environment. But the physical environment itself is the fundamental part of the field. My analysis would be: (1) Physiography: a study of land forms, that is physical, or geographic, geology; (2) Meteorology and oceanography, meteorology being quite as fundamental as the study of land forms in determining life conditions; (3) Biogeography, a study of ecology, that is the response of living things, plants and animals to the physical environment (1 and 2) and the consequent distribution of forms; (4) Human ecology, a study of the geographic conditions of human culture. This would include political and commercial and military and some other phases of geography. The fifth term in the series passes beyond geography, is the field for which geography should be the conscious and purposeful preparation, economics, civics, and sociology, yes and history too. I like to think of sociology as the fruit and flower of geographic study and that this service will prove the validity of the point of view of the geography of today."

As a matter of intellectual priority, Goode's statement is of great significance, for he may well have been the first American geographer to have adopted the term, "human ecology." As a practical matter, his clearly organized viewpoint was of consequence in establishing geography at Chicago. It does seem as though it was by accretion, and not by synthesis, that Goode derived and then exploited geographic insights obtained from Williams Morris Davis and Emory R. Johnson. The development of the Department of Geography at Chicago was a vitally significant achievement, and provided impetus for the formation of a scholarly society.

Geographical Societies in 1903

In 1903 it was not possible to reach a consensus of scholarly geographical thought, because the machinery to make a competitive discussion possible did not exist. To be sure, the nineteenth century had witnessed an increase of popular interest in scientific societies, including many that were called "geographical societies." But the support for these societies came very largely from business people, teachers, government workers, explorers, military personnel, and educated people with an unspecialized interest in science and exploration. The subject matter presented in these groups frequently related to exotic foreign lands and peoples, and at home to problems of settlement beyond the frontier. By century's end, exploration of the polar world had become a major interest and concern. Subject matter was intrinsically interesting to memberships, but study was without benefit of discipline. Even so, the work and accomplishments of these societies contributed towards the development of geography as a scholarly field. Developments abroad were not without significance for geography and geographers in the United States.

Meanwhile, Philadelphia and Boston were the two American cities where attendance at lectures on scientific and philosophical questions had become very popular. The American Philosophical Society was formed in 1743 at Philadelphia under the leadership of Benjamin Franklin. In 1859 Louis Agassiz and Asa Gray debated the concepts of evolutionary change in organisms at the Cambridge Scientific Club and the American Academy in Boston. Audiences came to listen, however, not to discuss. The most serious limitation of the effectiveness of these debates was "the lack of a comprehending audience."[34] The American Association for the Advancement of Science was formed in 1848, patterned after the British Association. In both organizations, Section E was devoted to Geology and Geography. But because most of those who attended the meetings were geologists, there was little opportunity to discuss geographical questions. Other societies fostering a geographical interest included the Geographical Society of the Pacific (1881), the Geographical Society of California (1891), the Geographical Club of Philadelphia (1891: renamed the Geographical Society of Philadelphia, 1897), the Agassiz Association (1892), the American Alpine Club (1893), the Geographical Society of Chicago (1898), the Peary Arctic Club (1899), and the Geographical Society of Baltimore (1902). The two United States geographical societies of major significance were the American Geographical Society (New York), and the National Geographic Society (Washington, D.C.).

The American Geographical Society[35]

The American Geographical and Statistical Society was formed in New York between 1851 and 1852, by a "group of wealthy philanthropists," whose objective was to provide a source of geographical and statistical information about parts of the United States that were not well known, and also about foreign countries where American merchants were transacting business. The first lecture before the Society, on November 8, 1851, given by Asa Whitney, was on a proposed route for a railroad to the Pacific. In the 1850s most of the lectures dealt with various aspects of the westward movement. In 1871 a revised charter omitted the words "and statistical," not, according to John K. Wright, because of a lack of interest in statistical information, but to make the name of the Society less cumbersome. The Society built one of the world's finest libraries of geographical books (classified geographically) and periodicals, and also a very valuable map collection. Over the years the American Geographical Society became a major center of geographic research, supported by gifts from individuals, or by grants from foundations, or by government contracts. In 1852 the Society began publishing a geographical periodical, which for many decades was the only such periodical in the United States. The *Bulletin of the American Geographical Society*—with some changes of its name—was published without interruption from 1852 to 1915, when it was succeeded by the *Geographical Review.* The American Geographical Society made possible some very important first steps toward the development of the field of geography in the United States. In the early years of the present century, however, its membership, and especially its officers, were less interested in the establishment of a new field of learning than they were in maintaining their own institution and providing practical services to American business and government.

The National Geographic Society[36]

The National Geographic Society was formed in Washington in 1888 by some thirty-three men who were prominent in Washington scientific and literary circles. Among the founders were Gardiner G. Hubbard (elected first president) and several of the men who had guided and worked on surveys of the American West after 1869. Prominent among these were John Wesley Powell, then Director of the U.S. Geological Survey and Chief of the Bureau of Ethnology; Henry Gannett, Chief Geographer of the U.S Geological Survey; C. Hart Merriam, Chief of the U.S. Biological Survey; Clarence E. Dutton, a geologist on the staff of the U.S. Geological Survey; and Adolphus W. Greely, arctic explorer who had been named Chief of the Weather Bureau in 1887. The Society initially had a membership of scholars interested in a variety of geographical studies. The first monograph, published in 1895, included three substantial scholarly . papers:[37] William Morris Davis on "The Physical Geography of Southern New England," Bailey Willis (of Stanford University) on "The Northern Appalachians," and John Wesley Powell on "Physiographic Regions of the United States." This was the kind of forum that Davis wanted to create, located in the midst of the largest concentration of scholars devoted to the study of geography in its diversity that existed in North America at that time.

By 1898 the Society was, however, experiencing financial difficulty. The membership included scholars rather than wealthy businessmen, and scholars could not pay for the cost of publishing the magazine and the monograph series. Efforts to increase the number of subscribers to the magazine failed. In 1898, Alexander Graham Bell (inventor of the telephone) became the second President of the Society. He proposed to popularize the magazine and to accept for publication only those articles written in simple, non-technical language, which avoided complicated controversy. A magazine with this kind of content, he said, would attract a much larger group of readers—large enough to cover the cost of publication. He looked for just the right person to be editor of this popular geographical writing, and selected Gilbert H. Grosvenor, a recent graduate of Amherst College.[38] Grosvenor soon eliminated the kind of scholarly and technical papers that had been written by Davis and others because, as he said, they were not intelligible to the average reader. Grosvenor, who was named Editor-in-Chief of the Society's publications in 1903, sought to create a new popular image of the Society. The membership increased from 1,417 in 1899, to 2,500 a year later, to 3,500 in 1904, and to 337,446 in 1914.

Meanwhile, Davis and those who supported his concept of a society of scholars were concerned by this departure of the National Geographic Society. In 1902 the creation of a new class of members to be known as Fellows was suggested, to which only those who had made original contributions to geography would be elected. The separate membership of "Fellow" was established by the Society, but never instituted, for as Bell said it was a "class distinction. . .all very well in a monarchical country. . .but somewhat out of place in a republic like the United States."[39]

In 1899 a crucial event had taken place on the occasion of the Seventh International Geographical Congress in Berlin. Without consulting other geographical societies in the United States, Bell wrote a letter, and asked the Hon. Andrew D. White (United States Ambassador to Germany) to present it to Ferdinand von Richthofen, President of the seventh Congress. Bell's letter invited the next Congress, scheduled for 1903, to meet in Washington, and proposed sponsorship by the National Geographic Society. The letter, written September 15, 1899, reads in part as follows:[40]

> The National Geographic Society, with an active membership of about 1,200 and 800 corresponding members, distributed in all parts of the United States, embraces the great majority of working and teaching geographers in America; it publishes a leading geographical journal, and is closely identified with geographic teaching and the more important geographic research of the country.

When Baron von Richthofen accepted the invitation, Gilbert Grosvenor published an article in *The National Geographic Magazine,* entitled "Next International Geographical Congress to be held in Washington."[41] Although Grosvenor mentioned the possibility of collaboration with the American Geographical Society, the Appalachian Mountain Club of Boston, and the geographical societies in Philadelphia, Chicago, San Francisco, and Seattle, it appears that such possible collaboration was discussed only after Richthofen's acceptance. The Council of the American Geographical Society had wished to sponsor the ninth Congress in New York, and suggested St. Petersburg, Russia, as the venue of the eighth. When the National Geographic Society went ahead with the invitation for the eighth Congress, relations between the two largest geographical societies in the United States became strained. This was in part relieved early in 1903 when Adams, of the American Geographical Society, and representatives of other geographical societies journeyed to Washington, D.C.

to attend a meeting called by the National Geographic Society to discuss plans for the Congress. Davis, who had attended the Berlin Congress as a representative of the National Geographic Society, and who was chairman of the Committee on the Scientific Program for the impending eighth Congress in Washington, recognized the differences of the established societies. He determined then to found a learned society of mature geographic scholars.

3

The Formation of the Association of
American Geographers, 1903-1904

As Chairman of the Scientific Committee for the Eighth International Geographical Congress in 1904 William Morris Davis sought papers by American geographers. Geography in the United States was on display. Davis was uncertain, however, how many "mature scholars," who had made original contributions to some branch of geography, there were in the United States. He was well acquainted with those geographers who had been working in physical geography, and he had formed an opinion about the scholarly qualifications of people such as Ellen Churchill Semple and Martha Krug Genthe, both of whom had published studies in the *Bulletin of the American Geographical Society* and the *Journal of Geography*. He was less well acquainted with economic geographers, trained by Emory R. Johnson. The largest concentration of geographical scholars at the time was in Washington, D.C., where field men such as John Wesley Powell and Grove Karl Gilbert, had helped organize the National Geographic Society, in 1888. But Powell had retired to Maine in 1897 and had died in 1902. To make the best possible showing at the Congress, Davis realized that it would be necessary to mobilize the small number of American geographic scholars. Support of the Eighth International Geographical Congress and the formation of a scholarly society was essentially part of the larger problem of mobilizing geographers in the United States.

The American Association for the Advancement of Science Meeting in St. Louis, 1903

In 1903 Davis was elected a vice-president of the American Association for the Advancement of Science. The Association was organized in nine sections, each devoted to a field of scholarship, and each chaired by one of the nine vice-presidents. Section E was designated as Geology and Geography. Davis took the opportunity to address the attendees of this section on the need for more professionalism in geography.[1] He pointed out that although geology and geography were supposed to have equal status in Section E, no vice-president in the preceding twenty years had ever mentioned geography. Geology had already attained a respected position in the world of scholarship, in part through the focus of geological thinking provided by the Geological Society of America, founded in 1888.[2] Geographers were also in need of a professional society, which would provide a forum for the exchange of ideas, and where the results of research studies could be communicated to other specialists in the field. With recent changes in the National Geographic Society in mind, Davis urged the need for restricting membership to persons with a record of published original research in some branch of geography. He felt that it was not possible to conduct a meaningful geographic discussion in the presence of persons who had little or no professional interest in the field. The very large popular interest in "Geography" was a major reason for setting high standards of scholarship as a requirement for election to his anticipated geographers guild. Davis, a very practical person, stated:[3]

> The various recommendations that I have made are likely to remain in the air, or at most to secure response from isolated individual students, unless those who believe that the adoption of these recommendations would promote the scientific study of geography are willing to give something of their time and thought toward organizing a society of geographical experts—an American Geographers Union. From such a union I am sure that geography would gain strength.

Additionally he[4]

> outlined certain important conditions, namely; a suitable definition marking the boundaries of the subject; members to include those with whom geography was primary, or at least of not less than secondary interest, with a good measure of published work; and, "The independence of the union thus constituted, of all other geographical societies."

Davis then suggested who might be eligible for membership in a geographers' learned society, urging inclusion of geologists who had concentrated their studies on the last chapter of geological history, where interrelationships between man and his natural surroundings had been found. He also believed that some geography teachers in normal schools would be eligible for election on the basis of published writings. There were, he suggested, numerous potential members working in government agencies, such as the Geological Survey, state or national weather services, or in agencies dealing with hydrography, biology, ethnography, or statistics.

Mobilizing Support for Two Projects

Immediately after his address to Section E, "a preliminary meeting of a small number of geographers was held. . . . In this meeting the formation of a geographical association was advocated. The sense of the meeting was that steps should be taken toward the forming of such an organization." Thirteen persons assured Davis of their enthusiastic support. The thirteen were:[5] Charles C. Adams, Henry C. Cowles, John F. Crowell, Charles R. Dryer, Nevin M. Fenneman, Frederic P. Gulliver, Christopher W. Hall, Mark S.W. Jefferson, Curtis F. Marbut, W J McGee, Rollin D Salisbury, George B. Shattuck, Ralph S. Tarr. Four other geographers "expressed themselves in favor of the plan" though not in St. Louis—Henry G. Bryant, Angelo Heilprin, Marius C. Campbell, and Richard E. Dodge.[6] When Davis returned to Cambridge he continued to seek names of additional geographers, and whenever possible to hold small impromptu meetings to discuss both the International Geographical Congress and the formation of a professional American association of geographers. On January 26, 1904 he sent a circular letter to 17 persons already on his list, and to 15 additional geographers, whose names he had recently secured. The circular read:[7]

A private meeting for organization is proposed in connection with the Eighth International Geographic Congress to be held in Washington next September. (A circular concerning the Congress has been sent to you.) In the meantime it is requested that no public mention be made of the plan.

Will you be good enough to send to the undersigned: Any suggestions that you may wish to make concerning the formation of the Club, qualifications of members, action that should be taken before and at the Washington meeting, and plans for subsequent meetings; also names and addresses of persons in any branch of geographic work whom you would suggest as members of the Club, with some statement of their official position and published works.

(It is probable that a class of Fellows will be established by the National Geographic Society of Washington, only those who are regarded as geographical experts being eligible for this distinction. How does this proposed action by the N.G.S. affect your opinion as to the formation of the Club here proposed?)

It was provisionally agreed by those who discussed the matter at St. Louis that the first effort of the Club should be to promote the success of the International Geographic Congress which convenes in Washington, September 8, 1904, by individual action to that end. Allow me, therefore, to put the following question: Will you take membership in the Congress? Will you attend some or all of its sessions? Will you contribute one or more papers? Will you join the western excursion? Will you induce others to take similar part in the Congress? Will you try to secure several members for the Congress who may not attend its meetings, especially libraries and societies as well as individuals?

It was further proposed that a second meeting of the Club should be held in connection with the meeting of the American Association in Philadelphia, December, 1904. Is it probable that you can attend that meeting?

Some thirty people were invited to attend the "second preliminary meeting" in Washington D.C. at the time the International Geographical Congress was convening in that city. The proposal to establish a society for geographers was given enthusiastic endorsement "and it was resolved to proceed in a more formal manner toward the formation of a geographical association." A Committee on Organization was appointed to draw up plans for the first meeting which, Davis suggested, should be held in Philadelphia the following December during the meetings of the American Association for the Advancement of Science. Davis was named chairman of the Committee on Organization, which also included Henry C. Cowles (a plant ecologist at Chicago), Henry Gannett (Geographer on the U.S Geological Survey), Angelo Heilprin (geographer and explorer, founder of the Geographical Society of Philadelphia, then recently appointed Lecturer in Physical Geography at the Sheffield Scientific School, Yale University), and William F. Libbey Jr. (Professor of Physical Geography and Geology at Princeton University).[8] "This committee enlarged somewhat the list of original members, secured the nomination of a number of new members, prepared a draft of a constitution, and called a meeting for December 29-30 in Philadelphia. . ." The committee was also charged with the responsibility of drawing up a slate of officers for the following year, and of arranging a program. Davis informally drew up a list of subjects and authors:[9]

Subject	Author
Ocean Currents	Littlehales
Waves and Tides	Harris
Islands	

Mountains	Willis
Boundaries	Gannett
Deserts	Gilbert?
Plains and Plateaus	Davis?
Shore Lines	Shaler?
Waterfalls	
Caverns	
Valleys	
Deltas	
Cities	
Roads	
Rainfall	Ward or Henry
Winds and Storms	Rotch or Clayton
Volcanoes	Jaggar or Hovey
Glaciers	Reid

Davis also had in mind further unification of the field when he proposed to the American Geographical Society that a "League of American Geographical Societies" be formed as part of the plan to secure an "American Geographers Association." [10] But the Society rejected this notion and nothing further came of the idea.

The Eighth International Geographical Congress

The Eighth International Geographical Congress opened in Washington on September 8, 1904. President of the Congress was the distinguished polar explorer, Commander Robert E. Peary. Grove Karl Gilbert was acting President of the National Geographic Society, in the absence of Alexander Graham Bell. In addition to the National Geographic Society the Congress was sponsored by the American Geographical Society of New York, the Geographical Society of Philadelphia, and the Geographical Society of Chicago. Seventy-five foreign geographers attended the Congress, representing Great Britain, France, Germany, Austria-Hungary, Canada, Latin America, and Japan. One hundred and twenty-five papers were submitted, only a few less than the number submitted for the seventh Congress, held in Berlin in 1899. The United States Government Printing Office published 148 of the papers which had been presented at the Congress. [11]

The Congress remained in session from September 8 to September 22. Sight-seeing in the Washington vicinity, on the Hudson River, at Niagara Falls, and in St. Louis, and paper sessions at the

American Geographical Society, Niagara Falls, the University of Chicago, and the World Congress of Arts and Sciences in St. Louis, constituted the essentials of the Eighth International Geographical Congress.

It was, for many American geographers, their first meeting with international confreres. Friendships were formed, points of view exchanged, and correspondences established. American geographers realized that they had much to learn from their European colleagues. Davis took the opportunity to hold a meeting of American geographers interested in forming an "American Geographers Association."

The Selection of the Original Members of the Association of American Geographers[12]

Between the close of the International Geographical Congress and the first meeting of the Association of American Geographers, in Philadelphia on December 29-30, 1904, the Committee on Organization was confronted with the task of selecting "original members."

Seventy names were advanced as prospective members, and published papers of each candidate had to be evaluated to judge whether they did or did not provide evidence of "mature scholarship." Of the seventy persons proposed for membership, the committee rejected twelve. One of those rejected was J. Russell Smith, Emory R. Johnson's hard-working assistant on the Panama Canal study, and a student of Ratzel. There was no evidence, committee members believed, that Smith had been adequately trained in physical geography. Emory R. Johnson, however, was passed for membership. Of the fifty-eight persons approved as original members, ten declined the invitation to become members.[13] Most of the ten were in sympathy with the objectives of the organization, but were reluctant to join for various reasons, such as living too far from likely meeting places, the cost of the annual dues (set at $5.00), or because of too many other professional commitments. Two who declined membership joined later. Ten of the twelve persons who had been rejected by the committee, were admitted later.[14] The forty-eight original members of the Association included:[15]

Cleveland Abbe, Jr. (1872-1934)
Charles C. Adams (1873-1955)
Cyrus C. Adams (1849-1928)
Oscar P. Austin (1847-1933)
Robert L. Barrett (1871-1969)

Louis A. Bauer (1865-1932)
Albert Perry Brigham (1855-1932)
Alfred H. Brooks (1871-1924)
Henry G. Bryant (1859-1932)
Marius R. Campbell (1858-1940)
Frederic E. Clements (1874-1945)
Henry C. Cowles (1869-1939)
John F. Crowell (1857-1931)
Reginald A. Daly (1871-1957)
Nelson H. Darton (1865-1948)
William Morris Davis (1850-1934)
Richard E. Dodge (1868-1952)
Charles Redway Dryer (1851-1927)
Nevin M. Fenneman (1865-1945)
Henry Gannett (1846-1914)
Martha Krug Genthe (1871-1945)
Grove Karl Gilbert (1843-1918)
J. Paul Goode (1862-1932)
Herbert E. Gregory (1869-1952)
Frederic Putnam Gulliver (1865-1919)
Christopher W. Hall (1845-1911)
Rollin A. Harris (1863-1918)
Angelo Heilprin (1853-1907)
Robert T. Hill (1858-1941)
Ellsworth Huntington (1876-1947)
Mark S.W. Jefferson (1863-1949)
Emory R. Johnson (1864-1950)
William Libbey, Jr. (1855-1927)
George W. Littlehales (1860-1943)
Curtis F. Marbut (1863-1935)
François E. Matthes (1874-1948)
W J McGee (1853-1912)
C(linton) Hart Merriam (1855-1942)
Raphael W. Pumpelly (1837-1923)
Harry Fielding Reid (1859-1944)
William W. Rockhill (1854-1914)
Rollin D Salisbury (1858-1922)
Ellen Churchill Semple (1863-1932)
George B. Shattuck (1869-1934)
Leonhard Stejneger (1851-1943)
Ralph S. Tarr (1864-1912)
Robert DeCourcy Ward (1867-1931)
Bailey Willis (1857-1949)

Three other names had been proposed for membership, but their papers were not completed in time to be considered by the committee: all three became members the next year. The three who might have been original members were Wallace W. Atwood (Department of Geology, Chicago), Joel A. Allen (Curator of Mammalogy and Ornithology, American Museum of Natural History), and Arnold E. Ortmann (Curator of Invertebrate Zoology and Paleontology, Carnegie Museum, Pittsburgh, Pennsylvania).

The original members came with a wide variety of training and experience. The only common denominator was a somewhat vague concern with the study of the earth as the home of man. Only one person had earned a Ph.D. degree in geography: Martha Krug Genthe, the first student to complete the doctorate at Heidelberg with Alfred Hettner, then teaching at a girl's school in Hartford, Connecticut. However, both Emory R. Johnson and J. Paul Goode held doctorates which were essentially geographic, though earned in a department of economics. A career analysis of the original forty-eight members reveals that nineteen held positions as geologists, four could be called field geographers and explorers, and nine taught geography in secondary schools, teacher-training schools, or universities. In addition there were three ecologists, two oceanographers, one climatologist, one geophysicist, three economists and statisticians, two biologists, one ethnologist, and one diplomat. One was editor of a geographical periodical (Cyrus C. Adams), and one historian, Ellen C. Semple, who had studied with Friedrich Ratzel, at Leipzig, and was introducing Ratzel's ideas on anthropogeography to American thought. Twenty-two of the original members held positions in universities and teachers' colleges; sixteen were employed by the federal government; and the other ten held a variety of positions, or were self-employed. Among the forty-eight original members, fifteen had studied with Davis at Harvard. At least seven had taken some, or all, of their graduate training in European universities.

Most of the original members were middle-aged men. Only two women were included, Martha Krug Genthe and Ellen Churchill Semple. The majority (35 of the 48) were between 30 and 49 years of age. Only 12 were over 50, and only one, Ellsworth Huntington, was under 30.

Philadelphia, December 29-30, 1904

The Association of American Geographers was founded in Room 16, College Hall, University of Pennsylvania, Philadelphia, on Thursday, December 29, 1904. On that occasion, 26 geographers attended, of whom 21 were charter members. Interestingly, of the five geographers present who were non-members, four later became presidents of the Association: Douglas W. Johnson, 1928; Isaiah Bowman, 1931; Wallace W. Atwood, 1934; and J. Russell Smith, 1942. William Morris Davis, as chairman of the organizing committee, presided over the meeting, and Albert Perry Brigham, of Colgate University, was appointed Secretary, *pro tem.* The first business was a discussion of the draft of a constitution. The name of the organization was at once changed from the proposed "American Geographers Association" to "The Association of American Geographers."[16] Other minor amendments were made in the "Proposed Constitution," which was adopted unanimously. By this act the name of the society was made official, The Association of American Geographers.[17] The object of the Association, as stated in the Constitution:

> . . . .shall be the cultivation of the scientific study of geography in all its branches, especially by promoting acquaintance, intercourse, and discussion among its members, by encouraging and aiding geographical exploration and research, by assisting the publication of geographical essays, by developing better conditions for the study of geography in schools, colleges, and universities, and by cooperating with other societies in the development of an intelligent interest in geography among the people of North America.

The Secretary's report of this meeting includes a fuller explanation of what the organizing committee had in mind. "It is our desire," the Secretary wrote,[18]

> "to bring together the investigating geographers of the country. . . . While full membership is limited to those who have already published original work, every encouragement should be given to those who may be expected to publish original studies in the near future. The meetings of the Association are to be open to all interested persons, and those who have not yet been elected to membership may present papers on the program if they are sponsored by a member."

The amended Constitution provided that "the officers of the Association shall be a president, two vice-presidents, a secretary, a treasurer (one person may be both secretary and treasurer), and three councillors. These officers make up the Council, which shall manage the affairs of the Association." A slate of officers, nominated by the organizing committee, was elected for the coming year: President, William Morris Davis; Vice-Presidents, Grove Karl Gilbert and Angelo Heilprin; Secretary and Treasurer, Albert Perry Brigham;

Councillors, Cyrus C. Adams, Ralph S. Tarr, and Henry C. Cowles.

The organizing committee announced to the meeting that no regular publication would be recommended in the immediate future. It was agreed that the existing journals would afford sufficient opportunity for the publication of geographical papers. Customarily, the annual meeting of the Association would be held at the same time, and in the same place, as meetings of the American Association for the Advancement of Science, but the next scheduled meeting would probably be held in New York City. The Secretary's report also included the interesting comment that "a summer field meeting is in consideration." Apparently Davis had thought of a field meeting at the Delaware Water Gap in the summer of 1905, but criticism of this scheme on grounds of expense from mid-western geographers discouraged the plan.[19]

A letter that Davis had received from Emmanuel de Martonne of the University of Rennes, France, was read to the members. De Martonne reported on "the probable formation of a similar organization of European geographers." The members voted to send a letter to de Martonne, offering the best wishes of the Association of American Geographers, for the formation of a European association and cordially extending "the privilege of temporary membership with their American colleagues, should they have occasion to attend any of the meetings of the Association in America."[20] President William Morris Davis then read a paper, entitled "The Opportunity for the Association of American Geographers," in which he hoped to associate[21] "the students of the organic and inorganic sides, the human, economic, zoological, botanical, climatic, oceanographic, and geologic sides of geography. . .leading them to work in view of and in co-operation with each other, and to present their results in each other's presence, we shall have taken an important step in the development of geographical science; for it cannot be doubted that students on the different sides of our subject have as a rule lived too far apart."

Twenty-two papers were scheduled for the first meeting of the Association. . .19 by members and three "by invitation," that is, by non-members, each sponsored by a member. Owing to insufficient time on the two day program, 13 papers were read in full, and nine were read by title. Of the 22 papers, 13 might be considered "physical geography," four biogeography, two cartography, and one each was offered on transportation, literature, and earth grid. Authors and paper titles of this first meeting were:[22]

Bailey Willis, "Some Physical Aspects of China."

Frederic E. Clements, "The Interaction of Physiography and Plant Successions in the Rocky Mountains." Read by title.

Ellsworth Huntington, "The Seistan Depression in Eastern Persia."

Leonhard Stejneger, "The Distribution of the Discoglossoid Toads, in the Light of Ancient Land Connections."

Albert Perry Brigham, "The Development of the Great Roads Across the Appalachians."

Raphael W. Pumpelly, "Physiography of the Northern Pamir," (by invitation).

Ralph S. Tarr, "Some Instances of Moderate Glacial Erosion."

Douglas W. Johnson, "The Distribution of Fresh-Water Faunas as Evidence of Drainage Modifications."

Henry C. Cowles, "The Relation of Physiographic Ecology to Geography."

Reginald A. Daly, "The General Accordance of Summit Levels in a High Mountain Region: The Fact and Its Significance."

Isaiah Bowman, "Partly Submerged Islands in Lake Erie," (by invitation), Read by title.

Cyrus C. Adams, "The Improvement of American Maps," Read by title.

Richard E. Dodge, "The Journal of Geography and Its Purpose," Read by title.

François E. Matthes, "The Study of River Flow."

Lewis G. Westgate, "The Geographic Features of the Twin Lakes District, Colorado," (by invitation).

Nelson H. Darton, "Geologic Expression in Contour Maps," Read by title.

Harry F. Reid, "The Forms of Glacier Ends," Read by title.

Frederic P. Gulliver, "Muskeget, a Complex Tombolo."

William Libbey, Jr. "The Physical Characters of the Jordan Valley," Read by title.

William Morris Davis, "A Chapter in the Geography of Pennsylvania," Read by title.

Grove Karl Gilbert, "Moulin Sculpture."

George W. Littlehales, "A New and Abridged Method of Finding the Locus of Geographical Position, and Simultaneously Therewith the True Bearing," Read by title.

The first meeting of the Council, attended by Davis, Gilbert, Tarr, and Brigham took place on the evening of December 29.[23] Eight persons were recommended for membership. Cyrus Adams and Henry

Gannett were appointed to a committee on the improvement of maps (this was the first Association committee), and President Davis was appointed a committee of one to "take up the question of state books with publishers." The Council empowered Secretary Albert Perry Brigham to write a letter of introduction for Bailey Willis, recently returned from China, who planned a stay in Europe "to study certain problems of mountain structure and form." A similar letter of introduction was provided by Davis for the Robert L. Barrett—Ellsworth Huntington expedition to interior Asia.

4

The First Twenty Years, 1904-1923:
A Question of Identity

The policy of the Association of limiting membership to those who had published substantial research contributions led to slow numerical growth. In 1905 it had been agreed that to elect a new member "three-fourths of votes must be affirmative": in 1906 "nine-tenths of all votes cast" were necessary for election. Between 1905 and 1923, 130 new members were elected. Of these, 14 declined the invitation. Additionally, death and resignation took its toll and membership increased slowly. In 1904, there were 48 members; by 1912, there were 76 members; by 1920, 115 members; and by 1923 there were 130 members. Only qualified people were sought. Charles R. Dryer wrote in 1924,[1] "The A.A.G. . . . . .is. . . .the only geographical society in the world, admission to which requires something more than two friends and a fee." Many years later Earl B. Shaw reflected,[2] "When I became a member of the A.A.G. . . .those. . . chosen as members considered acceptance as a real distinction. I still have my letter of notification. . . ."

The opportunity for discussion among qualified scholars had now been brought into existence. It was a major accomplishment. The next step was to bring about that discussion. To facilitate this, Davis had hoped to remove the barriers between scholars with different intellectual training and posture. The geologists, whose influence in these early years was of major importance in the development of geography, had been trained in physics, chemistry, biology, and careful methods of observation in the field. Economists, whose influence on the developing field of geography was much less widespread, but who did offer a very different intellectual point of view, had been trained in economics, commerce, and mathematics. They formulated generalizations of wide applicability and mathematical and statistical procedures which could be described as theory. For the first decade these two groups had little common ground. Economists were in a minority. Walter S. Tower resigned in 1910, and Emory R. Johnson in 1914. In 1913, economic geographer Avard Bishop was elected to membership, but he declined. There were some geographers, such as Charles R. Dryer, who turned to humanists for deeper insights into the behavior of human beings. Some, like Ellen Churchill Semple, made use of the historical method. Out of this variety of contesting viewpoints emerged early twentieth century American geography.

These years (1904-1923) also witnessed a major change in the training of young geographers. In this period, the Ph.D. degree began to replace the M.A. degree as the basic requirement for appointment to academic positions. After the first department of geography, offering advanced study to the doctorate, was established at Chicago in 1903, there was a steady increase of Chicago-trained scholars entering the field. The first Ph.D. in geography, granted at Chicago, was awarded to Frederick V. Emerson, in 1907. His dissertation was entitled "A Geographic Interpretation of New York City." Starting in 1914, the flow of students trained as geographers in graduate schools began to increase rapidly. According to Whittlesey, doctorates awarded in geography included:[3] 1893-1903: Pennsylvania 3, Johns Hopkins 3; 1904-1913: Pennsylvania 2, Cornell 2, Yale 2, Harvard 1, Chicago 1; 1914-1924: Chicago 13, Wisconsin 5, Pennsylvania 4, Clark 4, Harvard 3, Yale 3, Cornell 1, Columbia 1. In 1920, for the first time, all candidates for membership in the Association, passed by the Credentials Committee and nominated by the Council, had been trained in graduate school as geographers.

These years may be appropriately characterized as "the search for identity." For people educated in a variety of scholarly fields, who had sought association because of an ill-defined common interest in the earth as the home of man, perhaps the most critical and omnipresent issue concerned definition of the nature of geography. There were occasional papers relating to the content of geography, and sometimes round-table discussion of the subject, but, for the most part, the quest for comprehension of the field came from the juxtaposition of the annual variety of papers read at the end of December. Formal and informal discussion followed which might be continued for several weeks by correspondence. Arriving at a geographical point of view was necessarily an individual matter which did not seem to become an urgent professional need until the later twenties and thirties. At the outset, Davis and his thought were respectable, respected, authoritative, and perhaps unfortunately, authoritarian. Certainly Davisian thought dominated for some years.

The Davis Model

William Morris Davis had helped to form what might be regarded as the first paradigm in American geography. Presented in his presidential address, at the second meeting of the Association in 1905, he summarized his thesis regarding the identity of this field of scholarship:[4]

"... .any statement is of geographical quality if it contains a reasonable relationship between some inorganic element of the earth on which we live, acting as a control, and some element of the existence or growth or behavior or distribution of the earth's organic inhabitants, serving as a response; more briefly, some relation between an element of inorganic control, and one of organic response. The geographical quality of such a relation is all the more marked if the statement is presented in explanatory form. There is, indeed, in this idea of a causal or explanatory relationship the most definite, if not the only, unifying principle that I can find in geography. All the more reason, then, that the principle should be recognized and acted upon by those who have the fuller development of geographical science at heart."

Davis then discussed other proposals regarding the nature of geographical studies. German geographers, he said, identified geography as the study of location, or distribution, leaving the nature of things located, or distributed, to be studied by other sciences. This is what we might call regional geography, where things of diverse origin exist in areal groupings. But if geography is to be more than empirical description, its elements must be treated systematically. He rejected the idea that geography should be restricted to the inorganic side of the field. He clarified his ideas about the overlap of geography with other fields of study, and pointed out that such overlaps are inevitable, and, perhaps, even desirable. The diffusion of plants and animals is a fact of great significance in geographical studies, but in most cases the limits of diffusion are found to be physical. He agreed with Ratzel and Reclus that it is not possible to restrict the organic side of geography to human geography. Moreover, it was illogical to define geography as dealing only with the present day. In all geologic periods there have been geographies, and much illumination can be derived from the examination of sequences of change through time, and by the recognition of recurring cycles of change.

Davis exhorted his students and followers to provide the field of geography with a balanced development, balanced, that is, between

2. On the American Geographical Society Transcontinental Excursion of 1912

the physiographic side and the ontological side. Nevertheless, he himself contributed little to the treatment of organic distributions. He continued to experiment with methods of observing physical features in the field, and of reproducing his observations in words, landscape sketches, and block diagrams. He sought explanatory descriptions, making use of theoretical models of sequences of change in landforms. What he called the "geographical cycle" was an idealized sequence of landforms developed by running water as an upraised surface is worn down to a surface of slight relief, close to the base-level of drainage.

The Annual Meetings; Programs and Papers

The annual meetings of the Association provided the first opportunity in the history of American geography for a carefully selected group of intellectuals to convene for the sole purpose of advancing geographical science. Explorers, people posturing for social position, philanthropists, the curious, and travelers and tourists had been given short shrift. The organization promised scientific advance. Yet there were difficulties to be negotiated. Membership was restricted, and although painfully cautious and careful additions were made each year, death and resignation gave cause for concern. Reginald A. Daly, Martha Krug Genthe, Walter S. Tower, and Emory R. Johnson, were only some who resigned from the Association in the first decade. Angelo Heilprin, Willard D. Johnson, A. Lawrence Rotch, and Ralph S. Tarr were perhaps the most notable members to have died in the first decade. Nevin M. Fenneman wrote to Albert Perry Brigham in 1912:[5]

> Rotch's death is surely another blow, and only a little beyond 50. I have become so obsessed with a feeling of fatality that when the Titanic went down I rushed at once to my correspondence to see whether Davis might perhaps have been aboard.

And there were ideological problems. Although a majority of the Association founders were geologists by profession, there was a constant concern to have the Association maintain a geographical point of view. At the Baltimore meeting, in 1908, Grove Karl Gilbert, in his capacity as President, urged special consideration of the matter.[6] Herbert E. Gregory summed the matter in a letter to Brigham:[7]

> Don't you think it would be wise to replace Fenneman as Treasurer by a man who is nearer to the geographic type? I believe in Fenneman and like him very much, but I am becoming exercised over the fact that each year the official

staff of the Association consists chiefly of men who, to my mind, are geographers only by a stretch of that term. I do not consider Fenneman, or Salisbury, or myself, geographers and there is certainly no need of a special society to take care of men like us. If the organization is large, I see no reason why geologists with geographical leanings would not be enrolled as members; but I think that only rarely should they occupy positions as officers. . . .

You see from these remarks that I am a little bit worried over the outlook. Wouldn't it be better to put the Association pretty largely in the hands of a group of younger men who clearly are geographers,—the type of man who in a few years will constitute the Society? And would it not be wise for me and Fenneman and certain others who are pretty clearly geologists to resign from this organization, so as to make the cleavage between geology and geography even more distinct?

Gregory, Chairman of the Geology Department of Yale University, had there seen the problem in embryo—of a young and unestablished geography trying to make its way against an invested geology. As matters eventuated, the ontographic part of the subject developed swiftly, more geographers were trained in the universities, and by the twenties geologists and physiographers had become a minority group.

Initially, geologists had been welcomed into the Association, to give it numbers, stature, size, and intellectual strength. But once the Association became established, and a going concern, geologists could be "deferred," or "rejected," at time of membership application. Whom to include and whom to exclude became a contentious matter. It led to some embarrassments, divisions, and personal unhappiness and resentment. Yet this was the price that seemingly had to be paid if contributions and resultant discussion were to be of the highest calibre. This difficulty was exemplified by a letter Harlan H. Barrows wrote to Isaiah Bowman when the latter had requested support for his proposal of Robert Cushman Murphy, the distinguished ornithologist, for membership in the Association:[8]

I have believed for some time that the future of the Association is threatened by the continued admission to membership of geologists, botanists, zoologists, economists, and what not. Already the organization might well be called "The Association of Friends of American Geography" rather than The Association of American Geographers. I do not believe that animal ecology properly can be regarded as a branch of geography, and from my view point Mr. Murphy is a zoologist and not a geographer. In view of the stand I have taken on these matters, you will see that it would be embarrassing for me to second Mr. Murphy's nomination, and accordingly I am returning the blank to you.

On a card that Bowman attached to this letter he wrote:[9]

This is the sort of thing that fills me with despair! Are there not at least 30 or 40 members of AAG who are professed geographers (i.e. hold geographical positions, etc.) who have written less real geography than Murphy. Yet because he is a professed zoologist, he is not to enter. This is just too narrow for words. If AAG were an *Academy with real judges* Murphy would surely enter on first ballot.

Severe differences concerning an individual's nomination to Association membership could and sometimes did arise. Since ratification by 90 percent of the membership was required to admit an individual into the Association, those admitted were largely *en rapport*.[10] But it was among those who were "deferred," or "rejected," and the members who proposed those individuals, that feelings could and sometimes did run high.

The Association's annual program usually lasted two days, and was frequently scheduled to coincide with the meetings of the American Association for the Advancement of Science, and the Geological Society of America (between December 29 and January 2). Attendance, and the number of papers presented at the annual meeting of the Association, varied from year to year. For the first five meetings attendance averaged 25 members; for the years 1904-1923 attendance averaged 34 members. Between 1904 and 1913, 296 papers were delivered at the annual meetings of the Association, and in the years 1914-1923, 304 papers were delivered. This produced a twenty year average of thirty papers a meeting (worthy of note, however, is the bumper crop of 47 papers offered at the 1911 meeting).

It was initially suggested in 1904 that papers be delivered in approximately 15 minutes. By order of the Council, in 1906, it was agreed that delivery time for a paper should not exceed 15 minutes, and that discussion time for any member of the Association should be limited to five minutes. Even so, two days was not sufficient time in which to present all papers, and frequently several had to be "read by title." One of the two evenings of the annual conference was given to talks, sometimes by explorers, and sometimes by practising geographers: the other evening was usually given to a round-table discussion "smoker-style."

Individual contributions at these annual meetings warrant attention.[11] In the first ten years the most prolific contributors of papers were William Morris Davis (17), Ralph S. Tarr (12), Isaiah Bowman (9), Mark Jefferson (9), Lawrence Martin (9), Ellsworth Huntington (8), Albert Perry Brigham (7), Douglas W. Johnson (7), and Ray H. Whitbeck (7). In the second decade of Association history (1914-1923) the most prolific contributors of papers included Mark Jefferson (8),

J. Paul Goode (6), Ray H. Whitbeck (6), Albert Perry Brigham (5), and William Morris Davis (5).

There was, of course, a substantial offering in physiography, especially for the first ten years, though physiography retained vigor until 1923. Interesting was the departure from this physiography exemplified in several thrusts.[12] Studies in towns, population, and urban geography were advanced by Martha Krug Genthe, "Valley Towns of Connecticut," 1905; Mark Jefferson, "The Distribution of Population," 1907, "The Anthropography of Great Cities," 1908, "The Anthropography of North America," 1911, "The Growth of American Cities," 1913, "Regional Characters in the Growth of American Cities," 1914; Robert M. Brown, "City Growth and City Advertising," 1921; Marcel Aurousseau, "The Geographic Study of Population Groups," 1922. Regional geography was advanced by Isaiah Bowman, "Hogarth's 'The Nearer East' in Regional Geography," 1905, and "The Regional Geography of Long Island," 1909; William Morris Davis, "The Principles of Regional Exposition," 1913; Samuel Wiedman, "Northern Wisconsin," 1907; Lawrence Martin, "Lake Superior Region," 1907; George E. Condra, "Great Plains Region," 1907; and Walter S. Tower, "The Importance of Studies in Regional Geography," 1905. An embryonic physiological climatology coupled with climatic change in historic time began to emerge with Ellsworth Huntington, "Influence of Changes of Climate Upon History," 1906, "The Climate of the Historic Past," 1908, "The Big Trees of California as Recorders of Climatic Change," 1911, "The Effect of Barometric Variations upon Mental Activity," 1911, and "The Shifting of Climatic Zones as Illustrated in Mexico," 1912; J. Russell Smith, "The Origin of Civilization through Intermittency of Climatic Factors," 1908; and Robert DeC. Ward, "Climatic and Disease: How Are They Related?" 1905. Economic geography was advanced in rigorous manner by J. Russell Smith in "The Place of Economic Geography in Education," 1905, "A Proposed Division of North America into Economic Districts on Natural Regions," 1919, and, "The Division of North America into Economic Regions," 1921; Charles R. Dryer, "Natural Economic Regions," 1914, "Studies in Economic Geography: 1. Definitions and Classifications, 2. The Economic Regions of the United States," 1915; and Carl O. Sauer, "Geography as Regional Economics," 1920. Economic geography was to assume a larger role in Association programs in the thirties and forties.

In the first twenty years, there was also considerable investigation concerning the nature of geography. Specific studies concerning the structure of the field were worked out, methodology was studied, and established foreign orthodoxies examined. There was a very real need to define geography and its bounds, and to distinguish the subject from geology, or physiography, with "man" added as an after-thought. William Morris Davis was the unquestioned leader in this enterprise. Additional to his St. Louis address was his 1905 presidential address, "An Inductive Study of the Content of Geography," (1905), "Geography as Defined by Hettner," (1906), "Uniformity of Method in Geographic Instruction and Investigation," (1907, round-table), and "The Development of Geography in the United States," (1923). Charles R. Dryer offered some interesting thought in "Philosophical Geography," (1907), and a presidential address, "Genetic Geography: The Development of the Geographic Sense and Concept," (1919); Albert Perry Brigham offered "An Attempt at a General Classification of Geography," and "The Organic Side of Geography, Its Nature and Limits," (1909), and a presidential address, "Problems of Geographic Influence," (1914); J. Paul Goode presented "The Geographic Perspective in the Course of Study," (1906), "A College Course in Ontography," (1907); and Walter S. Tower read "The Problem of a Classification for Ontography," (1910).

Throughout the first twenty years of Association geography, ex-emplified in the presented papers, was the ubiquitous theme of the causal notion. William Morris Davis was its originator, but it was carried forward most energetically by Isaiah Bowman, Albert Perry Brigham, Ellsworth Huntington, Ellen Churchill Semple, and Ray H. Whitbeck. "Influence" gave way to "adjustment," and a variety of ill-defined "determinisms" emerged. Ontography was made the subject of intense study, and divided and then sub-divided. Exactly how, and in what proportions, man the actor should be studied in relationship to his environmental platform, became the center of intellectual debate. Papers presented at the annual meetings were frequently on the theme, or related to it. Other papers sought parameters to the types of claim which might be made. It was a theme which would not go away.

Each of these "geographies" found its own support, created its own posture, and learned to co-exist with other branches of the subject, in part as a result of the intellectual jostling which resulted from a reading, and criticism, of twenty years of geographical papers delivered before the Association. Here was a self-made, nativistic variety of American geography being hammered out on the anvil of Association annual meetings. Authors would begin to plan their papers for the following years soon after delivering the previous paper, and, if the paper were offered extemporaneously, it was usually after the most studied practice of extemporaneity. Such papers were borrowed and read, or copied, and then perhaps borrowed again. Frequently the paper, or a revised version, sooner or later appeared in print. It would be hard to overestimate the worth of the fellowship, and the splendid intellectual spirit of give and take which characterized these years.

Presidents and Presidential Addresses

During the first twenty years of Association history, eighteen presidents were elected. The apparent anomaly is explained by Davis's three terms: 1904, 1905, and 1909. Seventeen of the eighteen presidents were selected from the original founders. The only non-founding member elected to the presidency in the period 1904-1923 was Harlan H. Barrows (1922). Of the forty-eight founders, twenty-one were eventually elected to the presidency, and eight others were elected to the vice-presidency. Three men, who in the first twenty years were invited to the presidency, declined: Raphael Pumpelly, A. Lawrence Rotch, and Isaiah Bowman. Bowman was invited on a total of six occasions before accepting the office in 1931,[13] Ellen Churchill Semple was the first (and, to date, only) female president. Mark Jefferson was the first president elected from a normal school. Two presidents in these years considered themselves unworthy of the honor; Grove K. Gilbert felt he was too much the geologist to represent geographers,[14] and Henry G. Bryant felt he was too much a figure-head and, insufficiently, a scholar.[15]

To be selected, and elected, president of the Association in these years was a singular honor. The largest responsibility of the office was to prepare and deliver an address at the annual meeting. The addresses were usually published and ultimately were read by geographers in the United States and abroad, and by intellectuals from other disciplines. They represented geography in the United States. Because of this position of importance which they assumed, Association presidents devoted themselves to the task of writing their own essays in the most thoughtful and careful manner. The result was a series of addresses which nourished, nurtured, and advanced geographical thought.

Four of the addresses (1904-1923) were not completed. Angelo Heilprin died while in presidential office (1907). In 1909 William Morris Davis offered "The Italian Riviera Levante," but did not place it (in unaltered form) for publication. Rollin D Salisbury was undertaking field work in South America during the autumn of 1912 and wrote to Bowman that he could not return in time for the meeting. Henry G. Bryant, whose address at Princeton in 1914 was entitled "Government and Geography in the United States," never submitted a manuscript for publication. And Herbert E. Gregory, president of the 1920 Chicago meeting, read a paper entitled, "Geographic Basis of the Political Problems of the Pacific," but did not prepare a manuscript for publication. On the other hand, Robert DeC. Ward prepared, and had published, his 1917 presidential address, "Meteorology and War Flying: Some Practical Suggestions," even though the meeting had been cancelled owing to wartime travel restriction.

The addresses that were delivered and published include:[16] William Morris Davis, "The Opportunity for the Association of American Geographers" (1904) and "An Inductive Study of the Content of Geography" (1905); Cyrus C. Adams, "Some Phases of Future Geographical Work in America" (1906); Henry C. Cowles, "The Causes of Vegetational Cycles" (1910); Ralph S. Tarr, "The Glaciers of Alaska" (1911); Albert P. Brigham, "Problems of Geographic Influence" (1914); Richard E. Dodge, "Some Problems in Geographic Education, with Special Reference to Secondary Schools" (1915); Mark S.W. Jefferson, "Geographic Provinces of the United States" (1916); Nevin M. Fenneman, "The Circumference of Geography" (1918); Charles R. Dryer, "Genetic Geography: the Development of the Geographic Sense and Concept" (1919); Ellen Churchill Semple, "The Influence of Geographic Conditions upon Current Mediterranean Stock-Raising" (1921); Harlan H. Barrows, "Geography as Human Ecology" (1922); and Ellsworth Huntington, "Geography and Natural Selection: A Preliminary Study of the Origin and Development of Racial Character" (1923).

The addresses in the early years were similar in nature to the Davisian model. The address by Cowles, the very able botanist and plant ecologist, linked climatic and vegetation cycles. Tarr's Alaska presentation was quite Davisian. While Davis exploited the Grand Canyon and parts of the West as his physiographic laboratory, Tarr (together with Alfred H. Brooks and Lawrence Martin) exploited Alaska as his laboratory. In 1914 Brigham urged caution in the study of "influences of the environment," and was thereby the first Association president to examine critically, and to suggest revision of, the Davisian model. Mark Jefferson (1916) returned to the posture of Davisian physiography, although undertaking a task hitherto little attended. . .the definition of physiographic provinces in the United

States. Ward seems to have offered what was the first presidential address which, in large measure, offered practical advice.

These were the tentative movements of geographical posture which began to probe for alternatives to the Davisian model. The searching was not attributable to dissatisfaction with physiography, but was the inevitable evolution in geographic endeavor. Fenneman's often-quoted, searching address, of 1918, indicated something of the "Circumference of Geography," and grouped systematic specializations around a core of regional geography. It was followed, in 1919, by Charles Redway Dryer's innovative address, which was to become part of the chorological concept. Dryer's address elaborated upon a paper he had delivered on the occasion of the fourth Association meeting at Chicago in 1907: "Philosophical Geography: Strabo, Kant, and Bain." In his presidential address delivered at St. Louis, Dryer offered:[17]

> "It seems clear and beyond question that the psychological foundation of the geographic concept is the sense of distribution in terrestrial space. We must concede the pertinence of the doctrine of Kant that "geography is a narrative of occurrences which are coexistent in space." The idea, more sharply put by Bain in the statement that "the foundation of geography is the conception of occupied space," fits and includes every work generally recognized as geography from Strabo to Ritter and Reclus. With various additions and qualifications, it forms the essence of most of the current and accepted definitions of geography, of which quotation is unnecessary."

The address aroused little concern at that time. In part this may have been due to the fact that this was the first time the annual meeting had been held west of Chicago. Only 17 members and 21 non-members attended, providing a total attendance of 38 persons. Additionally, by 1919 there seemed to be more interest in practising geography than theorizing about it. Gregory (1920) and Semple (1921) offered addresses concerning the reach of "influence". In 1922, the little-published Barrows presented his now well known "Geography as Human Ecology." Perhaps Barrows derived this notion in part from J. Paul Goode, Henry C. Cowles, or perhaps from discussion with the on-campus sociologists. The address was regarded as distinctive but the subjugation of physiography implicit in the study of the relations between man and his natural environment was perhaps premature. Huntington offered a triadic causation of civilization and kept alive the omnipresent causal notion in his address (1923) concerning, "Geography and Natural Selection."

All told these published addresses totalled 251 printed pages. The shortest was two printed pages in length, the longest 35 pages, and the average nearly 17 pages. This was the enduring legacy of the Association presidents in the first 20 years.

Committees, Resolutions and Significant Association Decisions

In addition to papers and presidential addresses delivered at annual meetings, the Association advanced the cause of geography and geographers by committee work. From the inception of the Association Davis had been skeptical of the worth of committees:[18]

> I should like to see our Association run on lines involving the least work possible for committees and officers, and the greatest freedom for individuals. To this end, it would be interesting to have a report to show how many committees have been appointed since we set out, and what they have done. If you care to make up such a report, it might act as a wholesome discourager to future movers of committees.

Yet committees became an integral part of Association apparatus. They facilitated intellectual advances throughout the year, and most certainly accomplished ends that would otherwise not have been achieved. William Morris Davis was appointed (1904). . .a one-man committee "to take up the question of state books with publishers."[19] That same year Cyrus C. Adams and Henry Gannett were appointed a "Committee on Improvement of Maps."[20] Adams and Gannett corresponded and talked with publishers, and with map authors, and were successfully active. Consequently, a report of this committee in 1908 (made by Bowman on behalf of Adams) won acceptance and the committee was continued. Meanwhile in 1907 a subcommittee was established (Cyrus C. Adams, J. Paul Goode, and Ellen Churchill Semple) "to instruct on the proper use of the best maps."

The committee and sub-committee on maps was intimately related to another Association committee thrust—that on geographic education. In 1906, the first Association resolution was passed; "to set on foot lines of inquiry that. . . promote the betterment of geographical conditions in America."[21] In 1908 the problem of geographic education was confronted when Richard E. Dodge conducted a round-table conference on geography for secondary schools "strongly supported and marked by sustained interest."[22] Dodge had sent a questionnaire to teachers and high school superintendents. Based upon returns he was able to confirm that recommendatons of the Committee of Ten had been followed by very few schools. Difficulties arose because emphasis had been placed on physical rather than human geography. As Mark Jefferson put it in 1901, geography in the schools should deal with man on the earth—in that order, not the earth and man. Additionally a committee of five was established "To consider and recommend any needed changes in the teaching of secondary school geography."[23] The membership of this carefully chosen committee comprised Richard E. Dodge (chairman), Ralph S. Tarr, Ray H. Whitbeck, Curtis F. Marbut, and Albert Perry Brigham. Dodge reported on the committee's activity during 1909, and related the work of this committee to another formed in 1910 on a "State Survey of Educational Bulletins" (William Morris Davis, Richard E. Dodge, Nevin M. Fenneman, and Mark Jefferson) which did much to bring the subject of geography before school authorities in many states. In 1910 Rollin D Salisbury led a round-table conference ("a smoker") on "The Purposes of Geographic Instruction and the Phases of the Subject Best Adapted to these Purposes." The 35 members in attendance were disturbed by a reported decline of interest in physical geography in the secondary schools. After some discussion of the causes of this decline, the following resolution was passed:[24]

"That physical geography fully deserves to retain a place in the high school.

That the disappointment or dissatisfaction sometimes experienced regarding the results of teaching this subject is in large measure due to inefficient teaching. That as a means for removing this dissatisfaction superintendents and principals are urged to procure teachers of physical geography adequately prepared in this subject, and to entrust the subject only to such teachers.

That no teacher of physical geography should be appointed in any educational grade who has not made serious and special study of the subject in a higher educational grade.

These efforts coupled with presented papers and round-table discussions on the subject and an increased attention to geographic education in publication, were all part of the slow development of geography in the school system. Added impetus was received after the United States entered World War I and geographers were able to demonstrate their utility. But perhaps the most significant development with regard to geography in the schools came at the annual business meeting of the Association in 1914 (December 30) when a resolution was carried:[25] "It (the Association) welcomes with approval a movement to organize a National Association of Geography Teachers, and pledges to such an organization its hearty cooperation." Informal and practical advice was given by Association members to those individuals attempting to create what became known as the National Council of Geography Teachers. People teaching geography in grade schools, normal schools, colleges and universities, who were not eligible for admission to the Association, could join the National Council. Dodge placed the entire matter of geography in the schools before the membership (and, upon publication, a wider audience) with his presidential address "Some Problems in Geographic Education, with Special Reference to Secondary Schools."[26] From this time onward educational matters were largely transferred to the National Council of Geography Teachers.

Another Association committee, the work of which was highly significant, was the Committee on Physiographic Provinces, whose task was the "delimitation of physiographic provinces." At Chicago in December 1914, the Association devoted a session to the study of regions. The following April (1915), at the joint meeting of the Association and the American Geographical Society, the Association Council empowered President Richard E. Dodge to select committee members. Dodge selected Nevin M. Fenneman (Chairman), Marius R. Campbell, Douglas W. Johnson, François E. Matthes, and Eliot Blackwelder.[27] The committee reviewed a contending variety of "geographic" and "physiographic" provinces, and "natural regions" that had been mapped and otherwise written about. W.L.G. Joerg had summarized the literature in 1914,[28] as did Fenneman, who additionally published a map then acclaimed as the most comprehensive of its kind extant[29] (Fenneman's study was reprinted by the Association and sold as its first "separate"). The matter of "the provinces" was one of considerable interest to the membership, and not all were anxious for the establishment of an orthodoxy in the matter. Curtis F. Marbut wrote to Secretary Bowman:[30]

> Uniformity in official definitions, where they concern matters of detail and simple units, is necessary. Where broader relations are concerned when questions of general relationships are concerned I see no more reason for dictating uniformity than to require uniformity of plan in all buildings.

Later Marbut wrote to Bowman concerning the Committee:[31]

> The chief matter to be discussed by the Association. . .is the meaning of the term Physiography. A committee could restrict it to topography alone, to topography and rainfall, topography and temperature, topography rainfall and temperature, and many other factors. We must decide that in order to anchor the committee. . .

In 1921 Armin K. Lobeck and Curtis F. Marbut were added to the committee. During its life it generated much discussion at annual meetings and much correspondence. Fenneman's books on the physiographic provinces of the eastern and the western United States derived much from the exchanges developed by this committee.[32]

Other committees established included: Committee on the Spelling of Foreign Geographic Names for American Use (1921, J. Paul Goode, Chairman); Committee on Normal School Geography (1921, Robert M. Brown, Chairman); Advisory Committee on Agricultural Geography (1922, Carl O. Sauer, Chairman); Committee on Geographic Illustrations (1923, Carl O. Sauer, Chairman); and a Committee on Tidal Work (1923, no appointments recorded).

The Founding of the Annals and Support from the American Geographical Society

Since the founding of the Association, there had been a concern as how best to publish results of discussion, papers read, and presidential addresses delivered. The first evidence of this had come in 1905, when William Morris Davis, Cyrus C. Adams, and Albert Perry Brigham were appointed a committee to meet with a representative of the American Geographical Society, to discuss the possibility of devoting the February issue of the Society's *Bulletin* to the proceedings of the Association meeting.[33] Davis expressed reservatons concerning the wisdom of creating an Association publication. To Secretary Brigham he wrote:[34]

> I was surprised at the strength of feeling for an independent publication, and am not converted by it. How vastly better to have our proceedings go forth widely in the A.G.S. Bulletin, with no trouble or expense to us to address and post, than to have to supervise publication ourselves to get out a small edition (we could not afford a large one). . .

In 1909 a Committee on Future Policy was created (Richard E. Dodge, Mark Jefferson, and Nevin M. Fenneman) to consider the possibility of publishing the proceedings without increasing dues, or alternatively, dropping dues altogether and not commencing a publication. The following year (1910), Dodge reported from the committee that a publication was thought best.[35] Cowles, Brigham and Fenneman were appointed a committee to nominate a publishing committee. The committee on publication which was selected included Richard E. Dodge (Chairman), Henry C. Cowles, Alfred H. Brooks, and Ralph S. Tarr. This committee was authorized to spend $650; $100 for committee expenses, $50 for postage, and $500 for publication costs. There was concern among Association members that the publication would incur a deficit. To remove the fear, Nevin M. Fenneman pledged $25 a year for three years,[36] William Morris Davis offered $50 for 1911, 1912, and 1913 against any deficit properly incurred,[37] Mark Jefferson offered $25 within a period of three months,[38] and Alfred H. Brooks offered $25 within the next three years.[39] These gestures were encouraging, and the *Annals* was established as the official publication of the Association. The first volume (which carried the insignia of the Association designed by W.L.G. Joerg)[40] appeared as a single issue in 1911, and contained a summary of each of the annual programs dating from 1904. Each succeeding volume appeared as a single issue until 1922, after which the *Annals* was published quarterly. It was at the Association meeting in New Haven, 1912, that discussion by the membership provided guidelines for the type of article which should be published. Only Association members might publish in the *Annals,* and, at this time, "foreigners" could not be Association members. Inevitably, publication costs became a problem.

Cooperation with the American Geographical Society

Association dues, levied and retained since 1904, were five dollars per member per year. Sixty dollars purchased a life membership. Costs were minimal, and payment by members was assumed. Annual Association expenditure prior to 1911 did not exceed $100. In that year the Association began publishing the *Annals,* and deficits began to appear. At this critical time in the development of a geographic profession, the American Geographical Society extended its support. William Morris Davis, who was in New York City early in January 1913, met with Archer M. Huntington of the Society, who outlined a plan of cooperation between the two geographical organizations. Davis listed what he felt might evolve as advantages and disadvantages. On January 17, he sent a long "confidential" letter on the subject to Dodge, Brigham, Fenneman, and Bowman, requesting their assessment of the proposition. Shortly thereafter a formal proposal was made by the American Geographical Society of which the following is a resume:[41]

> The Society and the Association are to maintain each its own organization entirely independent of the other.
>
> In the Spring of each year the Society and the Association shall call a joint meeting in New York devoted to geography, to which the members of both organizations are to be invited.

The arrangement of the programs shall be in the hands of the Association.

The meeting place and the expenses of the meeting shall be arranged and paid for by the Society.

The first meeting shall be scheduled for April 3 and 4, 1914.

The Association shall continue publication of its *Annals.* The title page shall state that it is published by the Association, with the collaboration of the Society.

The editing, proof reading, and management of subscriptions and collections, shall be in the hands of the Association.

The expense of manufacturing shall be paid by the Society.

A copy of the *Annals* shall be mailed by the Society to each of its members, and a copy of the Society's *Bulletin* to each of the members of the Association, free of cost.

Returns from the sale of the *Annals* shall be collected by the Association and paid to the Society at such intervals as may be agreed upon.

All sums received by the Society in this way shall be paid into a special account devoted to geographical research and exploration, which fund shall be open to receive contributions from other sources.

Appropriations from this fund shall be made only in aid of such projects as may be approved by a joint committee of the two organizations, and any expedition aided from the fund shall be stated to be "under the auspices of the American Geographical Society and the Association of American Geographers."

Whenever the Association shall desire to hold a meeting of its members in New York, the Society shall make available any room or rooms in its own building not otherwise committed.

This agreement may be terminated by either party on giving six months notice to the other.

Reaction was favorable, and by April 16 the matter had been placed before the Council of the Association. President Henry G. Bryant wrote to Mr. Huntington of "a unanimous sentiment in favor of accepting the general plan of co-operation."[42] While reserving ratification *in toto* for the membership at the forthcoming annual meeting, President Bryant did agree to "the issuing of the *Annals* under the joint auspices of the two organizations. . .As to the question of the joint spring meeting to be held annually in New York City. . .the members of the Council in their replies expressed a virtual assent." The expense of printing the *Annals* was borne by the American Geographical Society, and, additionally, Association members henceforth received, free of charge, the monthly numbers of the *Bulletin of the American Geographical Society.* It was an altogether generous gesture by Mr. Huntington—the largely unseen patron of American geography in the first half of the twentieth century. Isaiah Bowman, acting secretary of the Association, drafted a final letter of agreement between the Association and the American Geographical Society, and forwarded it to President Henry G. Bryant, who in turn forwarded the letter to the Society.

The first joint meeting of the Association and the Society was held in April 1914. This was followed by meetings in the spring of 1915 and 1916. War intervened and the spring meetings were not resumed until April 1920, 1921, and 1922. Sessions were usually scheduled for Friday morning and afternoon, and Saturday morning. Organizers of these programs, frequently informally advised by William Morris Davis, invited a dozen or more speakers to present papers on particular themes, and frequently paid part, or whole, fare of the speaker. This ability to pay train fare, and, sometimes, part of the hotel bill for selected participants, strengthened the intellectual content of the meetings. It also facilitated invitation of guest speakers, such as Frederick Jackson Turner, who, in 1915, presented "Geographic Influences in American Political History." In sum, the first three joint meetings were characterized by a preponderance of physiographic papers, though there was an increasing number of papers dealing with other geographical matters.

When the joint meetings of the Association and the Society were resumed in 1920 some essential changes had taken place in American geography. College and university geography departments were being created, those already in existence were expanding, and the War had given an impetus to geographic study in grade schools. The influence of William Morris Davis (who had retired from Harvard in 1912) was declining; at the same time a younger generation of geographers endowed with a different viewpoint was emerging. The new generation of geographers was dominated by the University of Chicago. In 1919, of seven names advanced for nomination to membership in the Association, six, proposed by Chicago faculty, were accepted. Chicago had earned its place as a dominant force in American geography. Rollin D Salisbury, chairman of the department (1903-1919), had gathered together a fine collection of scholars, including J. Paul Goode, Harlan H. Barrows, Walter S. Tower, Charles C. Colby, and Wellington D. Jones. He also brought to the campus occasional lecturers including Ellen C. Semple, Bailey Willis, and Isaiah Bowman. From this department, with its emphasis on seminar discussion and field experience, there emerged a succession of graduate students who came to play significant roles in the history of American geography. Most notably perhaps these students included, Charles C. Colby, Vernor C. Finch, Wellington D. Jones, Carl O. Sauer, and Stephen S. Visher, all of whom were elected to membership in the Association in 1920.

At the last of the joint sessions, held in April, 1922, discussions of problems in the mapping and interpretation of land-use were introduced. Oliver E. Baker, Carl O. Sauer, and Hugh H. Bennett were part of the program, and all were keenly interested in these problems. Following a paper by Sauer, entitled, "The Problem of the Cut-Over Pine Lands of Michigan," a rigorous round-table discussion was held on "Methods and Problems in the Study of Land Utilization."[43]

It was immediately following this meeting that Isaiah Bowman, Director of the American Geographical Society, wrote to Richard E. Dodge, Secretary of the Association of American Geographers, giving six months notice announcing dissolution of the contract existing between the Society and the Association.[44] "The Society now feels that the Association is in a position to go forward with the publication of the *Annals* without further assistance." Bowman announced that the "joint research fund" amounting to nearly $4,000 would be given to the Association provided that the principal would be invested and the interest applied to publication of the *Annals*. Bowman gratuitously suggested annual Association dues of $10, and that by selling a further six or eight sets of the *Annals* each year, a total of $900 per year would be available, which would cover the cost of production of the *Annals*. In a further letter to Dodge, Bowman wrote:[45]

> "I believe that one of the highest purposes which the Association can serve at this time is the publication of papers embodying the results of original research. Should circumstances arise calling for extraordinary outlay in order to put out in adequate form some quite exceptional paper it would give us pleasure to consider the possibility of helping the Association. . ."

Is it possible that one reason for discontinuing the joint meetings in New York was the introduction of the new land utilization theme, quite different in purpose and results from the explanatory description of physical conditions along with organic responses to inorganic conditions? A new group of geographers had emerged who were deeply involved in practical programs of land-use leading to better land management procedures. The most active workers, in this new ad-

3. Joint meeting of the American Geographical Society and the Association of American Geographers, Courtyard of the AGS (probably April 1922)

vancing edge of geographic research, now needed their own conferences to discuss problems with which they were all personally concerned. As we shall see in the next chapter, the first of the Spring Field Conferences in the Midwest was held in May 1923, but the participants in these new meetings included none of those who had played roles in the joint conferences in New York. Only active field workers were invited to this new series of meetings.

The Makers of Association Policy

During these first two decades policy was created by officers as specified in the Constitution of the Association. The Constitution of 1904 stated that the officers "shall constitute a Council which shall manage the affairs of the Association. Nominations of officers were made by a committee of three members previously appointed for the purpose. . ." Five members could make independent nominations which were to be received by the Secretary thirty-five days before the annual meeting. Election of officers was by ballot at the annual meeting. The officers were to include a President, two Vice-Presidents, a Secretary, a Treasurer (one person could hold both positions as Secretary-Treasurer), and two Councillors. The number of Councillors was raised to three by the membership (1920), and it became customary for each Councillor to serve for a three-year term. The President was usually named as a Councillor after his service as President. In 1920, starting in 1921, the Constitution was amended to include five Councillors. Starting in 1911, the Editor of the *Annals* was also included as a non-voting member of the Council.

Responsibility for the guidance of policy, however, was not left solely with the Council. In those early years the secretary became a central figure in the guidance of policy. The secretary was selected for his responsibility, efficiency, and standing in the field. The first secretary, Albert Perry Brigham, in addition to possessing these qualities, was also well-liked, and served for an extended period of time, sufficient in which to develop an intimate acquaintance with the inner workings of the Association. Brigham had, moreover, been a student of Davis at Harvard, and held the greatest respect for the man's judgment and work. Between the years 1905 and 1912, Davis was twice president, and a councillor during each year when he was not president. Davis frequently wrote to Brigham and other Association officers of his thoughts for the next meeting—who should be encouraged to deliver a paper, and on what subject, and whatever else he thought might best be done to improve Association affairs. These suggestions, springing from a source of Davisian exuberance and enthusiasm, were invariably followed. In consequence, the guidance of policy was largely in his hands, and was faithfully carried out by Secretary Brigham. Yet, successful opposition to Davisian wishes had been accomplished on the matter of establishing the *Annals,* and on the matter of inviting explorers, and others whom Davis did not feel were scientists, to speak at the annual meetings. Even so, until 1913 there was little doubt that William Morris Davis held the authority to determine policy. Commencing in 1913, Davis no longer served as an officer. Yet he continued to exert considerable influence through letters which he sent to three of his former students, Albert Perry Brigham (Secretary, 1905-1913), Isaiah Bowman (Acting Secretary in Brigham's absence in 1912 and 1913, then Secretary until 1916), and Richard E. Dodge (Secretary in 1917, 1920-23). Oliver L. Fassig was Secretary 1918-1919.

Only four people served as Treasurer during this period. Brigham held both positions until he resigned as Treasurer in 1907. Nevin M. Fenneman was Treasurer from 1908 until 1912. He was followed by François E. Matthes from 1912 to 1919, and by George B. Roorbach, 1920 to 1923.

Following publication of the *Annals* in 1911, Richard E. Dodge was named Editor until 1914, and from 1916 to 1923.

In other words, control of the Association during these first two decades was largely retained by Davis and his disciples. The Secretary's influence was vital and pervasive. He prepared the agenda for Council meetings; he suggested names of persons who might serve on the committee to nominate officers for the coming year, or on the Credentials Committee which passed on the qualifications of proposed new members; he instructed both committees on procedures that had become customary; he received the titles and abstracts of papers offered for presentation at annual meetings; and he arranged the program, and supervised the local arrangements. Of course, the Secretary could not act as an arbitrary authority, for he always discussed policy questions with other officers, but he was in a position to make suggestions which were usually accepted and adopted.

One of the largest matters confronting the officers of the Association concerned distribution of the membership, venue for the annual meeting, and selection of officers. When J. Paul Goode was named chairman of the Nominating Committee in 1917, to select officers for the following year, he wrote to Bowman requesting advice. Bowman replied:[46]

"I wish I could give you assistance in the selection of men to carry on the work of the Association in 1918. I will not suggest the names of any individuals, but would suggest a policy—which is that we secure an entirely new set of officers from among the members of the Association heretofore inactive, or at least only moderately active. A glance through the list of members would furnish suggestions. Because if there is any feeling that the affairs of the Association have been in the hands of a small number we want to dispel that feeling. I had the same idea until I became Secretary and saw how unselfishly a small group were devoting themselves to the work of the Association and its affairs and thereafter I came to entertain a very high regard for them. It was wise to have control of the Association pass from the hands of Davis to others and it may be equally wise to have control go into the hands of a different group.

J. Paul Goode made a study of Association history, then wrote to Isaiah Bowman in 1917 stating that nearly 70 percent of the members resided in the eastern states (mostly in the Northeast and Middle Atlantic areas), 27 percent of the membership resided in the Middle West, and 3 percent were in the western half of the nation.[47] Goode also calculated that the East had provided 22 officers, the Middle West had provided 14 officers, and the West had provided none. Nearly 75 percent of the years in office had been served by the 22 officers from the East. Selection of the annual meeting place had clearly favored the East. Baltimore, Cambridge, New Haven, Philadelphia, Pittsburgh, and Princeton had each been selected for one meeting, Washington, D.C. had been selected twice, and New York City, three times. The only venue other than these eastern locations had been Chicago, which had sponsored the meeting twice.

This matter of "Eastern" dominance had been of concern to Association officers for some years. As early as 1905, Harold W. Fairbanks, of Berkeley, California, had urged a publication for the Association, for otherwise western members were totally divorced from Association affairs.[48] Later Fairbanks urged, to no avail, a meeting of the Association in California. Then he suggested that the annual meeting of the Association be made a part of the Panama Pacific Exposition. When nought came of these proposals, he wrote to Association Secretary Bowman in 1915 that the Association was "completely ignoring" the West. Bowman stated bluntly that cost of travel was the only problem, and then suggested to Fairbanks, "I do not know what others may think of a Pacific Coast branch, but I personally should much rather see a Pacific Coast Geographical Society."[49] In writing to Henry G. Bryant in 1913, of Association negotiations with the American Geographical Society, Albert Perry Brigham offered:[50]

Is there any danger that members of our Association. . .will feel that we are making too local an affiliation in New York? Will the Western members feel that we are becoming anchored too much in the East. . .as Secretary I have been anxious since the organization of the Association to avoid anything sectional, and to preserve equal interest and loyalty on the part of all members.

Following the annual meeting at Chicago in 1914, William Morris Davis wrote to Isaiah Bowman:[51]

Would it not be well, in view of eastern absentees at Chicago meeting, to organize an eastern and a central section of the A.A.G. to hold separate winter meetings for two years, and then every third year meet jointly on intermediate ground like Buffalo or Pittsburgh? Do no let us blink at the fact that distances are large and travelling expenses high. . .

Many years later, regional divisions were created.

5

The Search for Alternatives, 1924-1943

In the second twenty years of Association history significant changes occurred both in the content of American geography and in the distribution of American geographers. Several mid-western universities offering advanced programs of graduate geography were established, and others already in existence were strengthened. Many of the programs were staffed, in part, by geographers who had studied at the University of Chicago. The *Announcements* of the University of Chicago for 1933-34 noted that at the time the Department of Geography was represented by former students on the faculties of 130 other institutions of higher learning: universities, liberal arts colleges, and teachers colleges.[1] It was in these years that the Midwest began to rival the East as the seat of American geography. Notwithstanding the establishment of the Clark University Graduate School of Geography under the direction of Wallace W. Atwood (himself a one-time graduate student at the University of Chicago), control of the Association began to shift to the Midwest. Transfer of control of the Association, from the disciples of William Morris Davis to those of Rollin D Salisbury, was quietly and certainly one of the major developments of this period. This was reflected in the venues selected for the Association's annual meeting, and in the changing variety of geography offered on those occasions.

The Annual Meetings

Between 1924 and 1943 a total of 817 papers were delivered before the audiences of nineteen annual meetings (the 1942 meeting was cancelled owing to wartime travel restrictions). This was an average of 43 papers per meeting. Of these, approximately 62 percent were read by members and 38 percent by non-members.

Major contributors of papers were, in the decade 1924-1933: Wallace W. Atwood 10 (plus two co-authored), Mark Jefferson 10,

Derwent S. Whittlesey 10, Ray H. Whitbeck 8, Preston E. James 7, Glenn T. Trewartha 7, and Stephen S. Visher 7. From 1934 to 1943, the major contributors were: Stephen S. Visher (9), George B. Cressey (7), Richard J. Russell (7), H. Thompson Straw (7), Leonard S. Wilson (7), Wallace W. Atwood (6 plus 2 co-authored papers), Preston E. James (6), Robert S. Platt (6), Glenn T. Trewartha (6), Derwent S. Whittlesey (6), and Wallace W. Atwood, Jr. (5, plus three co-authored papers). Between them, these thirteen individuals provided 131 papers (and seven co-authorships).

Although the papers were written for the most part by acknowledged craftsmen, quantity cannot be regarded as the only index of intellectual excellence. There was widespread agreement during this period, among the generation of geographers then emerging from graduate school, that alternatives to the Davisian inspired model of environmental control and human response should be sought. A number of very able geographers had advanced thought by publishing really noteworthy contributions in the teens of the century. An impetus was given to these contending geographic varieties, and the dissemination of these works was, if not swift, enduring. Ellen C. Semple continued her fine elaboration of a causality derived from Ratzel; J. Russell Smith organized the field of economic geography and published what was probably the first text on the subject in the United States, *Industrial and Commercial Geography* (1913), then added *Commerce and Industry* (1916); Vernor C. Finch and Oliver E. Baker published *Geography of the World's Agriculture* (1917); Homer L. Shantz and Curtis F. Marbut thrust hard in the direction of vegetation and soil study; Ellsworth Huntington continued his search for the causation of civilization itself; Dodge and Bowman edited LeCompte's translation of Jean Brunhes' *La Geographie Humaine;* and a considerable number of geographers contributed to the war effort in a variety of ways (gathering data, constructing maps, writing reports for the Shipping Board or the Commodities Board, or working with the Military Intelligence Division).[2] Davisian physiography was still in evidence but plenty of alternatives were available. Physiography was no longer the orthodoxy, and the competitive discussion of geographical ideas and methods led to what might be referred to as the first "scientific revolution" in the history of American geography.

In the period 1924-1933, there was a restless searching for geographic objectives and methods. There was a widely held recognition that there were many "geographies." But there was also a sincerely held belief that, from this competing and contending tangle of points of view, there would emerge a synthesis. Meanwhile co-existence of the many differently held positions was not incompatible with a forward march of the geographic profession. Points of view began to develop around clusters of workers. The American Geographical Society, the Chicago group, the Berkeley group, the Minnesota group, and the Wisconsin group, were but examples. By 1932 the University of Wisconsin Geography Department had found it helpful to issue a "creed" concerning their working definition of geography. Bowman, Director of the American Geographical Society, was challenged repeatedly in meetings of the Social Science Research Council. In response to a variety of belittling remarks, he wrote *Geography in Relation to the Social Sciences* (1934). This questing for a comprehension of the structure of the field was the process of subject in quest of discipline. This was evident in the programs of the Association's annual meetings. Then into the ranks of professional geographers stepped Carl O. Sauer, a young man who had completed a doctorate at the University of Chicago and who had taken a position in the Department of Geology and Geography at the University of Michigan (1915). Sauer presented before the Association, "Economic problems of the Ozark Highlands of Missouri" (1919), "Geography as Regional Economics" (1920), "Problems of Land Classification" (1920, and published in full), and "Objectives of a Geographic Study" (1922). Other than these papers, he offered only three other presentations before the Association: "Memorial of Ruliff S. Holway" (1928), "Foreword to Historical Geography" (presidential address, 1940), and "The Education of a Geographer", (address as Honorary President, 1956). The impact of his first four papers was quite extraordinary. Future chroniclers of the history of geographic thought may well view this contribution as Kuhnian "extraordinary science."

Of "Geography as Regional Economics" Sauer wrote:[3]

There have been numerous discussions of the scope of geography and especially there have been examinations of the periphery of the science. Much less attention has been given to the determination of particular objectives within the field of geography. Geography is suffering from a scattering of interests over too broad a field for the limited number of workers engaged in it.

The focussing of attention on certain phases of the field alone appears to give hope of establishing the science solidly. This involves consideration of the content, aims, and methods of such a specific type of inquiry. In this country historical geography has been treated in such a manner.

> A voluntary limitation of research by a group of workers to the field of re-
> gional economic geography is probably the most urgent need of the science
> to-day. Regional economics has not been preempted by economic science
> and belongs most appropriately to geography. The essential problems are
> 1) the determination of bases of unity of the area, 2) the inquiry into advan-
> tages and handicaps inherent in the area, 3) the time element as affecting stage
> of development, and 4) the analysis of the entire economic complex of the
> region. It follows that any area, geographically defined, is an appropriate
> subject of inquiry, and that the inquiry must not be limited to the evaluation
> of so-called geographic factors.

> In the method of research, work needs to be done in forming a scientific disci-
> pline for 1) agrogeographic research, as referring to rural conditions, especial-
> ly the utilization of the land, 2) urban studies, and 3) movement of trade.

> A logical as well as pragmatic sanction is at hand for such studies, and by
> means of them geography may knock successfully at the door of the business
> world and, as well, it may present itself as an advisor to governmental policy.

This paper, coupled with Sauer's publication "The Problem of
Land Classification,"[4] seemed to provide a new direction for
American geography in the twenties. The purpose was practical, and
was applicable to programs for the improvement of the economy.
Shortly thereafter Sauer wrote *The Morphology of Landscape,*[5] and
opined that geography was the study of the associations and interconnections of things and events in areas. It was a publication of excep-
tional significance. Later, Sauer reflected,[6] "*The Morphology of
Landscape* was written in several weeks as a sort of habilitation
(because of the separation of geography from geology at Berkeley).
Also I was then emancipating myself from the direction then ruling at
Chicago that geography was a study of responses." Apparently he
"soon found the subject of the 'Morphology' distasteful, perhaps seeing that any restrictive definition hampers original work."[7] The
"Morphology" represented an ideological expositon and a logical extension of his thought delivered before the Association. His work
shifted the geography of American geographers.

Both Sauer and Wellington D. Jones had been invited to discuss
the land classification idea at the last joint meeting of the Association
and the American Geographical Society (April, 1922). The
geographers at this meeting were not particularly interested in field
surveys and associated technical problems. A need presented itself for
a special sort of meeting. At the annual Association meeting held at
Ann Arbor in December 1922, Jones and Sauer invited some of the
former participants in the Chicago seminar to a dinner. There they
drew up plans for an annual field conference to discuss problems and

4. *Carl O. Sauer (Photograph by Robert G. Bowman, 1937)*

5. One of many "Periodic Field Conferences" near Menominee, Michigan, June 1935. Front, left to right: Richard Hartshorne, K. C. McMurry, G. Donald Hudson, Charles C. Colby, Glenn T. Trewartha, Derwent S. Whittlesey. Rear, left to right: Fred B. Kniffen, W. L. G. Joerg, Samuel N. Dicken, Vernor C. Finch, Preston E. James, Stanley D. Dodge, Wellington D. Jones, Ralph H. Brown, Loyal Durand, Jr. In attendance but not in the photograph were J. Russell Whitaker and Robert S. Platt. Photograph from the Robert S. Platt collection.

experiences in land classification studies and to experiment with new procedures. It was resultant to this meeting that periodic field conferences were held from 1923 until 1940.[8] A second conference was organized in 1926 by junior members of the profession, mostly from midwestern universities. In 1935 the two "conferences" were combined.

Between 1924 and 1926 interest of the active minority of Association members was concentrated on field methods for land classification surveys. Geographers recognized the need to find better field methods by which the necessary information could be gathered and used for planning purposes. The matter of definition was held in abeyance. In 1924, 1925, and 1926 there were special sessions of the Association given to the discussion of field methods. These culminated in 1926 when Vernor C. Finch described the method he had adopted in mapping land use near Montford (Wisconsin).[9]

In the ten years following 1924 papers along the lines of this theme were delivered,[10] notably by Wellington D. Jones, Vernor C. Finch, Darrel H. Davis, Kenneth C. McMurry, Robert S. Platt, Derwent S. Whittlesey, Preston E. James, and Richard Hartshorne. Jones offered "Detailed Field Mapping in the Study of the Economic Geography of the Agricultural Areas" (1924), "Ratios in Regional Interpretation" (1928), "Field Mapping of Residential Areas in Metropolitan Chicago" (1930), and "Procedures in Regional Investigation" (1933). Finch contributed "A Detailed Map of an Agricultural Area" (1924), "Culture and Landscape at Madison, Wisconsin" (1925), "Progress in the Field of Mapping of Detailed Geographic Interrelationships" (1926), "The Service Area of Montfort: A Study of Landscape Types in Southwestern Wisconsin" (1928), "The Geographers Field Map as a Permanent Record of Landscape Forms" (1928), and "Reconnaissance Notes on the Geography of the Mississippi Delta Fringe" (1933). McMurry delivered "A Study in the Use of Soil Types in Geographic Mapping" (1925), "Work of the Michigan Land Economic Survey" (1926), "The Use of Land for Recreation" (1929), and "Aeroplane Mapping on Isle Royale" (1931). R.S. Platt read "A Classification of Manufactures, Exemplified by Porto Rican Industries" (1926), "Field Study of an Iron Range Community, Republic, Michigan" (1927), "Field Study of a Sugar District: Mariel, Cuba" (1928), "An Urban Field Study Marquette, Michigan" (1929), "Pirovano—Items in the Argentine Pattern of Terrene Occupancy" (1930), "Copper Mining Pattern of Terrene Occupancy in the South Range, Keweenaw Peninsula" (1931), "Mining Patterns of Occupancy in Five South American Districts" (1932), and "Terrene Occupance in the Maracaibo Basin" (1933).

This collection of papers represented a keen interest in classification and use of the land and other natural resources, which had developed in this group of midwestern geographers. It was an attempt to demonstrate the worth of planning, that is, planning with carefully gathered knowledge of the cultural and natural patterns combined. It was concerned with all the elements in the regional pattern, individually and in their collective association, revealing the composite pattern of occupance. Studies of small areas had been made by some who were not involved with state supported land classification surveys. In 1922, for example, Colby read "The California Raisin Industry",[11] which concerned a distinctive agricultural area near Fresno, California, and which for several years was adopted as a model for the geographic interpretation of the close adjustment of land use to the underlying characteristics of the land. The following year he offered another similarly detailed study of a small area in Nova Scotia, "A Geographical Analysis of the Apple Industry of the Annapolis Valley."[12] Whittlesey's notion of sequent occupance thrust the element of time alongside the spatial component of American geography in the twenties.[13]

In that decade, survey and classification had been the main undertaking of the geographers involved in this enterprise. By the conclusion of the twenties a new dimension—appraisal— had been added to this work. Strengths and weaknesses of the occupance pattern were sought. Studies in spacing of individual elements, and their relationship to the composite pattern were made, and the range of geographic complexity revealed. Two advances in technique came in the early thirties: aerial mosaics provided the map base, and site analyses were developed by geographers. Special mention should be accorded Kenneth C. McMurry, who presented a study (Ypsilanti 1931) of Isle Royale, a thinly populated island in Lake Superior which had been covered by overlapping vertical air photographs.[14] From these photographs he produced a mosaic of the island, and comparing observations made on personal traverses with the aerial photographs, he compiled a land-use and vegetation map. This study led to the replacement of the plane table and compass by a new and revolutionary technique for field mapping, so inaugurating a new era. Site analyses were made by a number of individuals, and by the geographical section of the Tennessee Valley Authority.

By 1935 Charles C. Colby was chairman of the sub-committee on land classification of the Land Committee of the National Resources Planning Board, and Harlan H. Barrows was a member of the Water Resources Committee of the National Resources Board. C. Warren

Thornthwaite was working for the Soil Conservation Service, which had been planned and administered by Hugh H. Bennett. G. Donald Hudson was senior geographer in the Department of Regional Planning Studies of the Tennessee Valley Authority. Curtis F. Marbut, Homer L. Shantz and Oliver E. Baker had demonstrated the utility of geographical planning in the field of agriculture. Isaiah Bowman developed the concept of planning in pioneer areas throughout the world.[15]

Emerging in the Association program, after 1933, was a regional geography. Regional papers, of course, had been read previously. Walter S. Tower had delivered "The Importance of Studies in Regional Geography," in 1905. Preston E. James wrote some of the earliest studies of small areas as regions. This owed, in part, to the intellectual ancestry of Richard T. Ely, who taught Oliver Baker. The latter transmitted the point of view of a sequence of change in economic patterns to Preston E. James who was studying at Clark University in 1922. James then applied this point of view to field work he prosecuted in Trinidad in 1924.[16] James followed this with "The Blackstone Valley: A Study in Chorography in Southern New England," and "Regional Planning in the Jackson Hole Country."[17] Beginning in the early thirties, however, the concept, purpose, and worth of regions received more attention than hitherto. The 1933 and 1934 meetings of the Association both devoted a session to a conference on regions.[18] In his (1935) presidential address, Charles C. Colby observed that "nearly every American geographer in his arguments or by his work has emphasized an interest in the study of the regions which make up the Earth. That we entertain varying points of view as to how to study regions should make for advances in our science."[19] This interest in the region had been fostered by studies of smaller areas. Some philosophical and methodological differences were apparent in annual programs, but a goodly number of substantive regional contributions were made. This was the time in American geography when the study of the region began to approach high art. Papers in regional geography were contributed most notably by:[20] George B. Cressey on China; Stanley D. Dodge on New England; Loyal Durand, Jr., and Vernor C. Finch on Wisconsin; Robert B. Hall on Japan; Roland Harper on Europe; Preston E. James on Latin America; Robert S. Platt on the United States and Latin America; Stephen S. Visher on Indiana; Ray H. Whitbeck on Latin America; and Derwent S. Whittlesey on the United States, Europe and Africa.

Papers in the fields of business were contributed by Helen M. Strong, and in economic geography more notably by Richard Hartshorne. Rural and urban morphology also began to assume a permanent place on the Association's annual program. Contributions to population geography and urban geography were made by Mark Jefferson, Clarence E. Batschelet, Henry M. Kendall, and Preston E. James, among many others. Political geography received definition and elaboration during the thirties with notable contributions by Richard Hartshorne, Lawrence Martin, and S. Whittemore Boggs. A special session entitled "The Changing Political Geography of the World" (Chicago, 1939), and other papers were inspired by the outbreak of hostilities in Europe. Cultural geography was likewise emerging as a genre, with papers most notably by Fred B. Kniffen and Lester E. Klimm. The polar world was frequently remembered by William H. Hobbs, W.L.G. Joerg, and W. Elmer Ekblaw.

William Morris Davis watched his infant creation grow, and he continued to display the keenest interest in the Association and attended meetings to present papers quite frequently. His last such offering, entitled, "Clear Lake, California," was in 1933. By that time the percentage of papers dealing with physical geography was lower than it had been at any time in the first twenty years of Association history. Leading contributors, other than Davis, included Wallace W. Atwood, Wallace W. Atwood, Jr., and Richard J. Russell.

The Presidential Addresses, 1924-1943 [21]

Of the twenty presidential addresses in this period eighteen were published; the address by François E. Matthes (1933) was not published and the scheduled address by W.L.G. Joerg (1937) was neither delivered nor published. Four presidential addresses were composed by charter members of the Association: Curtis F. Marbut (1924), J. Paul Goode (1926), Marius R. Campbell (1927), and François E. Matthes (1933). Each address represented the special interest of the respective president, and, collectively, they comprise a most valuable addition to the literature. The subject matter exhibited is remarkably diverse.

The only recurring theme seems to be that dealing with the history and substance of the field. Douglas W. Johnson delivered "The Geographic Prospect" (1928), Charles C. Colby offered "Changing Currents of Geographic Thought in America" (1935), W.L.G. Joerg prepared, but due to illness did not deliver, "Generalization and Synthesis in Geography" (1937), and Vernor C. Finch read "Geographical Science and Social Philosophy" (1938). Isaiah Bowman thought long and hard of delivering an address concerning the scope and purpose of geography, but then settled on "Planning in Pioneer Settlement" (1931).

Other presidential addresses in this period included: Curtis F. Marbut—"The Rise, Decline, and Revival of Malthusianism in Relation to Geography and the Character of soils" (1924); Ray H. Whitbeck—"Adjustments to Environment in South America: An Interply of Influences" (1925); J. Paul Goode—"The Map as a Record of Progress in Geography" (1926); Marius R. Campbell—"Geographic Terminology" (1927); Lawrence Martin—"The Michigan-Wisconsin Boundary Case in the Supreme Court of the United States, 1923-1926" (1929); Almon E. Parkins—"The Antebellum South: A Geographer's Interpretation" (1930); Oliver E. Baker—"Rural-Urban Migration and the National Welfare" (1932); François E. Matthes, "Our Greatest Mountain Range, The Sierra Nevada of California" (1933); Wallace W. Atwood, "The Increasing Significance of Geographic Conditions in the Growth of Nation States" (1934); William H. Hobbs—"The Progress of Discovery and Exploration within the Arctic Region" (1936); Claude H. Birdseye—"Stereoscopic Photographic Mapping" (1939); Carl O. Sauer—"Foreword to Historical Geography" (1940); Griffith Taylor—"Environment, Village and City: A Genetic Approach to Urban Geography, with Some Reference to Possibilism" (1941); J. Russell Smith—"Grassland and Farming as Factors in the Cyclical Development of Eurasian History" (1942, but delivered in 1943); Hugh H. Bennett—"Adjustment of Agriculture to its Environment" (1943).

This anthology of presidential addresses constitutes a rich trove indeed and bespeaks a diversity which led to a questioning of the nature of the field.

"The Geography of American Geographers"[22]

In 1930, and again in 1932, Almon E. Parkins sent a questionnaire to more than forty members of the Association, asking for their thought concerning the purpose of geography. From approximately thirty replies which Parkins received, he established five groups of definition: the relationship of man to his natural environment; a systematic study of the facts which can be studied "distributively"; human ecology; description of the face of the earth; ways in which mankind occupies and exploits regions of the world. Parkins concluded that "Geographers in America, no matter at what level they work, are certainly in general agreement that boundaries (but not barriers) to their science should be set up."

Isaiah Bowman shared a concern that geographers should make explicit the nature of their complex undertaking. He had grown weary of the division which accompanied the diversity of American geography, and he wished to see it more clearly defined. As Director of the American Geographical Society, and as a member of a number and variety of committees, many of which related to education. Bowman was aware that geography did not enjoy equal standing with other sciences in the United States. "As President of Social Science Abstracts Inc., and as a representative of geography on the Commission for the Study of Social Sciences in the Schools, and also as a member of the Social Science Research Council, I have some responsibility in this respect. . ." He wrote *Geography In Relation to the Social Sciences* (1934), a book relating to "the philosophy of geography as well as its place and value in scholarship and education."[23] The book presented Bowman's philosophy of what geography was, or should be. He had little patience with techniques, measurements, and, in fact, the work of the midwestern geography conferences. The book resolved little, but it was of especial use to teachers of geography and teachers of the social sciences. And it did provide workers in other fields with a notion of what geography and geographers hoped to achieve.

Four years later Richard Hartshorne's *The Nature of Geography* was published. Unlike Bowman's book, which amounted to a personal manifesto, Hartshorne's study was an excursus concerning what others had thought, spoken, and written about the nature of geography. The immediate origin of Hartshorne's monumental work was in two studies he had completed in 1934—the first, a talk given to the social studies division of the Minnesota Educational Association, on "Objectives in Teaching Geography"; the second was a paper entitled, "Recent Developments in Political Geography."[24] Both undertakings challenged Hartshorne to think of the content of the field. Some geographers in the United States believed they were following Sauer, when they challenged Hartshorne's work in political geography, in proclaiming that the subject matter of geography should be limited to material, physically observable, features. This view was presented by Glenn Trewartha at the Association meeting in Evanston in December 1933. A vigorous exchange of correspondence followed and Hartshorne came to realize "that what was necessary first was scholarly study of the sources." Carl O. Sauer and Richard Hartshorne were opposed on the matter of the position of political geography within the corpus geographic. Sauer followed Otto Schlüter and Siegfried Passarge. Hartshorne followed Alfred Hettner. At the Thirteenth Institute of the Norman Wait Harris Memorial Foundation at the University of Chicago, June 21-28, 1937, a conference was held on "Geographic Aspects of International Relations." Hartshorne recognized a marked lack of consensus among the geographers present as to the content of their field. "Looking back now to the academic year 1937-38. . .it appears that I was becoming programmed—again in no small part through the existence of the Association of American Geographers—to make a major effort to examine critically the thinking of American geographers about our subject."[25] At the Association meeting held in Ann Arbor, in December 1937, Hartshorne presented a paper dealing with the political geography of the Mid-Danube Basin. Trewartha again challenged the legitimacy of a study not dealing with physically observable objects. Hartshorne urged the rejection of any one authority, and later challenged "the cult of discipleship that seemed to have developed." Sauer had not attended an Association meeting for several years. The second incident at the 1937 Ann Arbor meeting was a reply, by Robert S. Platt, to an article by John B. Leighly, attacking concepts then current concerning regional geography, which had been published in the *Annals.* At a lunch meeting following presentation of this paper, discussion ensued and Hartshorne informed Whittlesey, who was then editor of the Association's *Annals,* "that the paper should have been consigned to the circular file." Whittlesey requested that Hartshorne state his criticism in a form that could be published. There began a study which, with a sojourn in Europe and encouragement from Whittlesey, evolved into *The Nature of Geography,* first published in two issues of the *Annals,* then in a single volume reprint. Revision and reprints later were to impose themselves again and again upon the author "in what is perhaps a never-ending struggle for intellectual integrity in our profession."[26]

Without the ambit of the Association, *The Nature of Geography* would not have been written.

Committees, Resolutions, and Significant Association Decisions

During the second twenty years of Association history, committees continued to play a significant role, though rigorous pruning of committees which had run their course may also be observed.[27] In 1924 the Council discharged the Committee on Normal School

Geography, the Committee on Geographic Education, the Advisory Committee on Agricultural Geography, and the Committee on Policy (appointed in 1922). No new committees were created. Continued were; Committee on Publication, Committee on the Delimitation of Geographic Regions, Committee on Geographic Illustrations, Committee on Tidal Work, Committee on Physiographic Boundaries and Provinces, Committee on the Spelling of Foreign Geographic Names for American Use, Committee on Finance. These were supplemented in 1927 with Committees on "Facilities for Geographic Research in Washington, D.C." and "Opportunities for Trained Geographers in Government Service" (both of these committees had been urged by Ray H. Whitbeck). In 1930 three further committees were established,—on library classification, "to secure the Incorporation of Maps of Minor Civil Divisions in the Census Reports, and to formulate a Statement as to Desirable Changes in Existing Copyright Laws and in Laws as to the Dating of Maps." Three one-man committees were established in 1933, . . .on Publicity, Funding for Publication of Research, and Incorporation of the Association. Guy-Harold Smith functioned well as publicity committee for the Association. He informed periodical editors of Association affairs, and made known the work of the Association to a much larger group than hitherto. Glenn T. Trewartha, in charge of the quest for funds to publish research, had a thankless task while the Depression lingered. Robert S. Platt initiated inquiry concerning incorporation of the Association, which was accomplished in 1937 by the committee then consisting of Claude H. Birdseye, R.H. Sargent, and Helen M. Strong. Throughout the remainder of the thirties, committees were few and usually established for the discharge of a particular task. In 1934 a committee was established "to investigate possibilities of making up a Bibliography of Regional Studies in the Geography of North America" and in 1936 a Committee on Population Mapping was formed.

A group of authors, under the editorship of Almon E. Parkins and J. Russell Whitaker, produced a book in 1936 entitled *Our Natural Resources and Their Conservation.* Royalties from this book were evenly divided between the Association and the National Council of Geography Teachers.[28] Meanwhile, Wallace W. Atwood had donated $500 to the Association to further the prosecution of research in physical geography. He contributed a further $2,500 for a similar purpose in 1936. Subsequent to these developments, the Association established a Committee on the Use of Research Funds (1938).

In these years the Association decided to involve itself in matters pertaining only to the science of geography. The membership list was distributed with care. The Association refused support for a variety of causes—teachers' labor unions, anti-narcotic leagues, state park associations, bird-life societies, publishers of books of travel and adventure, and a variety of political organizations. Only one resolution seems to have been made by the Association in these years, and that was ratified by the Council, not the membership. On March 17, 1936, Secretary Preston E. James dispatched the following letter to the President of the United States, to the Secretary of the Interior, and to the Chairman of the Senate and the House Committees on Public Lands:[29]

> This letter is to urge that no action be taken looking toward a modification of the present prohibition against entry or mineral appropriation within the Glacier Bay National Monument, Alaska, until an adequate examination of the area has been made by the United States Geological Survey.

The most perplexing and persistent matter to confront the Association in these years related to membership, and the question surrounding the reading of papers. Membership did not grow rapidly. That was ordained by the constitution. Only geographers who had made an original contribution to scholarship were eligible. In the twenties and, indeed, in the thirties, deaths and resignations were persistent, allowing only slight numerical growth in Association membership. Meanwhile departments of geography were growing in number and size in colleges and universities, and increasing numbers of geographers were obtaining doctorates. This was soon to create problems. The economic depression also hindered growth. In 1927 the Association comprised 136 members and by 1932 the number had dropped to 133. In the latter year, six members died, four of whom were among the original forty-eight members and past presidents: Albert Perry Brigham, Henry G. Bryant, J. Paul Goode, William G. Reed, Ellen C. Semple, and Frederick J. Turner. In 1933 a Committee on Nominations for Membership was established. The following year the word "credentials" was substituted for "nominations." Candidates' applications were studied carefully. Some candidates were "approved," some "deferred," and some were rejected. Feelings of nominator and nominee were frequently hurt by deferment or rejection. Some began to consider the Association an exclusive club. Because non-members could not publish in the *Annals,* it was felt by some that they did not have equal professional opportunity. This charge was not totally without foundation, for the Director of the American Geographical Society, whose geographical point of view was largely shared by the editor of *The Geographical Review,* may not have been well disposed towards the newly emerging land-use and land-survey viewpoints, and repeated discussion of details of technique in a small area.

Non-members could, however, be introduced at annual meetings of the Association, where they might read a paper, if time permitted. Some introduced papers were read by distinguished scholars from other fields, and occasionally scholars from abroad. Other individuals introduced were still working toward a doctorate, and this practice was frequently questioned. In 1928, Association Secretary, Charles C. Colby, recommended that a statement be drawn up concerning "title, character and presentation of papers for the annual meeting" which would be included in the letter concerning the call for titles for the annual meeting.[30] A committee was established on this matter, (Charles C. Colby, Marius R. Campbell, Ray H. Whitbeck) from which Colby as chairman reported in 1931, on "Suggestions on the Presentation of Papers Before the Association of American Geographers." Of "introduced papers" he wrote, "(they) have become the great problem of the program committee. . . It should be remembered that the purpose of the Association is to serve as the clearing house for ideas among mature geographers. It is not a training school for young geographers. It appears wise. . .to adopt the general practice of not introducing young men who are students in our graduate schools. . . The young man who can speak extemporaneously with sufficient skill to interest the AAG probably is unborn."[31] The ratio of non-members' to members' papers had been increasing steadily since 1928 when the ratio of non-members reading papers had been 16 to 32. In 1929 the ratio was 16 to 19; in 1930, 17 to 25; in 1931, 24 to 31; in 1932, 12 to 16; and in 1933, 20 to 30.

In 1936 the Committee on Programs (Richard E. Dodge, Wallace W. Atwood and Charles C. Colby) submitted a report to the Council imposing tighter controls upon both members and non-members who planned to read a paper at the annual meeting. The Council suggested that 75 percent of papers on the program should be presented by members and approximately 25 percent by non-members. The problem became acute in 1936, at the Syracuse meeting. The programs were held in a hotel, and were open to members and non-members alike. But the business meeting, which was open to members only, was held on the campus of Syracuse University. The Secretary (then Preston E. James) seeing the large number of non-members left at the hotel without a program, suggested to one of the non-members, Edward A. Ackerman, that he organize a round table discussion.[32] (This group met again in 1937 and 1938 at Association meetings and in 1939, at Chicago, formally announced the formation of a Young Geographers' Society.) In 1937 26 papers were presented by members and 25 by non-members. Secretary James reported to the Council, "deplored. . .is the smaller number of contributions from members—only 26 papers from 149 professional geographers. . ."[33] In 1938 the Secretary's report again referred to the matter, "There is, in general, a division of opinion between those who would limit the meetings to members only, and those who resent any attempt to reduce the number of introduced papers." The matter of membership restriction had become problematic. With the onset of World War II there were set in motion forces which ended the Association as it had previously existed. A new, larger, and more representative Association was to emerge.

The Council and Officers, 1924-1943

The Council had been enlarged to nine persons following constitutional revision in 1920, combining a president, one vice president, a secretary, a treasurer, three members of the Council (one to be elected each year for a three year term), and the two past presidents, each of whom was automatically appointed to the Council for a two-year term. Since 1911 the editor of the *Annals* had sat on the Council, but without privilege of vote.

At each annual meeting, a nominating committee was appointed by the Council to prepare a slate of officers to be elected by the membership at the next annual meeting. Usually the chairman of the committee drew up a list of eligible candidates and submitted it to the Secretary in June or July. The latter would send to the membership a ballot listing the nominees. The marked ballots were returned to the Secretary to be counted at the next meeting. Although there was provision for additional nomination, the candidates advanced by the nominating committee were, invariably, uncontested.

The nominating committees accepted their charge in most responsible manner. It was customary for the Secretary to instruct the chairman of the committee regarding the steps in the nomination process. He might pass on suggestions regarding candidates, especially for President, that had been made at the meetings of the periodic field conferences, where one evening was usually devoted to a discussion of Association affairs. Illustrative of the manner in which the chairman might present the problem to his committee is a letter from Almon E. Parkins, who was chairman of the nominating committee in 1934 to select officers for 1935. The other members of the committee were Lawrence Martin (Chief of the Map Division, Library of Congress), and Richard Hartshorne (University of Minnesota). The following letter, from Almon Parkins to his committee, was dated February 1, 1934:[34]

> Our committee is requested by the council to select the following officers for 1935: president, vice-president, treasurer, secretary, and one member of the council.
>
> No one will doubt, I am certain, that selecting a president calls for the most care and most thought by us. There is no *one* best man for the place, and therein lies our problem. We have so many eligibles and so many worthy ones that the only suggestion I can offer is that in selecting a president we should strive to maintain during a period of years, a fairly even balance in the "types of interest" represented in our membership. And if we find from a study of the list of past presidents we are off-balance we should strive to correct the balance. There is no secret that there are in our membership geomorphologists or physiographers, human geographers, and also a group of eminent men in related fields, as ecology, agricultural economics, soil science, etc. All, to repeat, should be represented in the list of presidents, and represented somewhat in proportion to the membership of each group in the Association.
>
> When the Association was young geomorphologists dominated in numbers (in fact they started the Association) and it was proper that the presidents should be chosen largely from this group. Of the first ten presidents, seven were physiographers, one was a plant ecologist, and two (Adams and Heilprin) I am unable to classify. In the past ten years five have been physiographers (including the 1934 president), four human geographers, and one an agricultural economist.
>
> I think there is no doubt that human geographers now outnumber either of the other groups in membership in the Association. This can readily be checked. Certainly they outnumber other groups at the annual meetings.
>
> We should also strive to maintain a rough balance as to geographic distribution.
>
> Moreover, the president and all other officers, in fact, should be active members in the Association.

Following exchanges of correspondence, the Committee selected Charles C. Colby as the nominee for President. The Committee then replaced Robert S. Platt (Treasurer since 1929) with John E. Orchard of Columbia University, since it was thought not good practice to have two officers from the same university (both Colby and Platt were at Chicago). Frank E. Williams of the University of Pennsylvania was continued as Secretary. The slate of officers that was unanimously approved was sent to Frank E. Williams on July 16, 1934.

Election to any office in the Association was regarded as a professional endorsement, and was doubtless of some utility in furthering a career. Office holders could regard themselves as having an immediate stake in the development of geography in the United States. The presidency was the highest office which the Association could bestow. The office of Secretary was most critical to its well being. In those days, the Association functioned pursuant to the work of the Secretary, who undertook this assignment additional to his livelihood. Charles C. Colby held the office from 1923 to 1928, Darrell H. Davis from 1929 to 1931, Frank E. Williams from 1932 to 1935, Preston E. James from 1936 to 1941, and Ralph H. Brown from 1942 to 1945.

The Council determined Association policy, ratified or rejected the work of individuals and committees, and otherwise assumed responsibility for the course of Association history. Decisions which the Council was obliged to make were usually of a similar nature from year to year. This changed in the early 1940's.

6

The American Society for Professional Geographers, 1943-1948, and the Association of American Geographers, 1944-1948: A Time of Change

The onset of World War II brought profound changes in American life. Learned professions were challenged in different ways. Some that could be related to the war effort were the beneficiaries of expansion, not necessarily along lines that had been foreseen. Geography was one of those fields.

As America's participation in the war defined more clearly the implication and extent of her commitment, certain of the agencies of government sought persons with the varieties of training and education that the study of geography provided. Large numbers of individuals were sought who were experienced in the design, construction, and use of maps, and who could attach significance to locations. During the late 1930s, some 40 geographers were in Washington, D.C.; by 1943 there were some 300 geographers employed there. The demand for geographers was, moreover, heightened across the country when an opportunity to introduce geography into college and university offerings through an enlarged Army Specialized Training Program presented itself.

Isaiah Bowman had discussed the role of geography in the Army Specialized Training Program with Colonel Herman Beukema (Director of the Army Specialized Training Division) and Secretary of War Stimson. He had also discussed the subject with numerous other officials in the Pentagon. He wrote a confidential note to Gladys Wrigley: "I have been in Washington two days, as a member of a committee of nine, setting up the curricula to be taken in large doses and in quick succession by the 150,000 who will be sent to the colleges for specialized training. Our two-day session was intense, and I want to tell you one day what I did with the section on geography!"

Bowman's advice to Colonel Beukema, establishing a plan and a program for the creation of an appropriate geographic literature and sufficient geographic teaching, may have been decisive in adding the subject matter of geography to the Army Specialized Training Program.[1]

So it was that geographers and surrogates were hired to teach geography. This heightened the competition for geographers' services. Many of the geographers already in Washington were obliged to work intensely, as there simply were not enough geographers to meet all war-time requirements. Many of these geographers were younger people who had not yet had time to produce "an ample record of published studies." Some, however, were mature scholars who had devoted themselves to the daily round of university and college teaching to the neglect of research studies. Consequently, many of the geographers in Washington were not eligible for membership in the Association. The concentration of so many geographers in one city was unprecedented. It is calculated that by September 1943 there were 217 professional geographers employed in more than 20 different agencies, and the number increased as the months passed. There were 75 geographers in the Research and Analysis Branch of the Office of Strategic Services (initially known as the Coordinator of Information), 46 in the War Department, 23 in the Intelligence Division (G2), and an additional 23 in the Army Map Service. The Office of the Geographer, Department of State, had 13 geographers; 15 were employed in the Board on Geographic Names; 12 in the Office of Economic Warfare; 12 within the Department of Agriculture. Additionally, eight geographers were employed by the Geological Survey, and six by the Coast and Geodetic Survey. There were five geographers in the Weather Bureau, four in the Map Division of the Library of Congress, and eighteen others were scattered among a variety of agencies (these figures do not include draftsmen or others engaged in producing maps and charts). Additionally there were approximately 25 geographers in posts overseas.[2]

Among all these, only 45 were members of the Association. By December 15, 1941, the total AAG membership was only 167. Many influential members of the geographic profession had, seemingly, not met the requirements for membership in the Association of American Geographers. In a summary of the situation existing in 1943, Chauncy Harris noted that of 32 living graduates with doctorates from the University of Wisconsin, only 13 had won election to the Association.[3] Membership requirements were still essentially those of 1904. Those elected frequently felt they had earned the "honorific," and were reluctant to lower standards. However, a number of geographers who had emerged from colleges and universities felt disenfranchised. Some

91

The American Society for Professional Geographers, 1943-1948, and
The Association of American Geographers, 1944-1948: A Time of Change

were bitter, referring to Association members as "super Ph.D.'s" and "elitist." For these individuals the Young Geographers was the only alternative to membership in the Association.

At this critical juncture of the early 1940's, Association officials were both scattered and distant from Washington, D.C. The Secretary, who assumed office following the New York annual meeting, December 1941, was Ralph H. Brown of the University of Minnesota. It was at this time that Brown was put in charge of the geographical work in the "Language and Area Program" of the Army Specialized Training Program. Brown not only offered courses himself, but also was obliged to provide a hurried program of guidance for non-geographers assigned to teach geography.[4] Treasurer of the Association was Guy-Harold Smith, of the Ohio State University in Columbus, Ohio. Editor of the *Annals* was Derwent Whittlesey of Harvard University. J. Russell Smith of Columbia University in New York City, was President in 1942 and Vice President was Nels A. Bengtson of the University of Nebraska. All of these men were distant from Washington, D.C., and not fully aware of the intensity of feeling which could and did develop there.

Ralph Brown organized the thirty-ninth annual meeting which was to be held in Columbus, Ohio in December 1942. It had been planned as a joint meeting with the American Historical Association, with special sessions devoted to the relations between geography and history, a matter in which Brown was personally interested. In the autumn of 1942, however, the meeting was postponed, at the request of the National Transportation Authority. Consequently, the thirty-ninth and fortieth meetings were held jointly in September 1943 in Washington. J. Russell Smith delivered as his presidential address "Grassland and Farmland as Factors in the Cyclical Development of Eurasian History."[5] Hugh H. Bennett's presidential address, "Adjustment of Agriculture to its Environment,"[6] was delivered on the second day. This Washington meeting was attended by 225 geographers, the largest meeting of the Association to that time.

A New Professional Society

Yet all was not well with American geography and geographers. There was developing an ever-increasing dissatisfaction among a number of competent and enthusiastic geographers who had not been admitted to the Association. One of the earliest expressions of a desire to have these geographers participate came from Clyde F. Kohn, then President of the Young Geographers Society:

James has written me saying that the Council of the A.A.G. vetoed his proposal to make us associate members of that organization. As a result: Do you think that we should make the Young Geographers' Society a more definite organization with a more effective form of membership. . . . Such a plan you recall was vetoed at our Chicago session in 1939. However I feel that such an organization might be worthwhile.

William Van Royen had also wished to change the existing order, but in a more fundamental way than that proposed by Clyde F. Kohn. He wished to exact changes in the Constitution and by-laws of the Association, and to that end he circulated a letter to a number of geographers. Replies came quickly: from Wallace W. Atwood,[8] "I fear your suggestions for changes in the constitution and by-laws of the AAG are too revolutionary. . .We are all in sympathy with the general purpose of your suggestions to invite many more of the younger members of our profession into the Association;" from J. Russell Smith,[9] "I am very glad to hear that ideas are boiling in the minds of yourself and others. How would it do for some of us to get together informally at the Cosmos Club and have a powwow?. . ."; from Stephen B. Jones,[10] "After reading the constitution and your proposals, I feel more strongly than ever that it is running headlong into difficulty to tackle the problem that way. The constitution merely states that 'membership shall be limited to persons who have done original work in some branch of geography.' All that we really need to do is to moderate some of the practices that have grown up about that clause." Thought and concern was given to Van Royen's proposal of change, but the Young Geographers Society remained the only alternative to membership in the Association. Early in the spring of 1943, F. Webster McBryde, Head of the Latin America Section, Topographic Branch, Military Intelligence Service of the War Department, and his assistant, George F. Deasy, attended a meeting of the Young Geographers. Deasy, a competent and effective research geographer, was displeased with the AAG. Younger of the two, he had at that time published little of a scholarly nature to qualify him for membership in the Association, but his confidential war-work was substantial and of a high order. He very much needed a forum and an active professional group. His resentment at exclusion from the Association was shared by many other geographers. It was Deasy who persuaded McBryde to go with him to the meeting. McBryde, four years his senior, was the possessor of extensive field experience and scholarly publication and his election to the Association appeared imminent. These two men were dissatisfied with the meeting. Fifty or sixty, mostly younger men, sat listening to a talk by the chairman of their group, Shannon McCune, himself a respected scholar, teacher, and authority on the geography of Korea. McCune urged these professional geographers to contain themselves, assuring them that his mentor, George B. Cressey, had promised that if they would be patient they would all eventually be admitted into the Association.[11] But the Young Geographers realized that with the restrictive membership policy of the Association, decades would have to pass before they could all win admission. Additionally, since the work of many of these geographers related to classified war-related government projects, "publication" had to remain of a confidential nature.

McCune was retiring as chairman, and elections were planned for new officers of the Young Geographers. Deasy persuaded McBryde to run for the office of chairman. McBryde agreed to run for the office provided that if elected he would be given free reign to establish a national professional society of geographers independent of the Association. The one other candidate for president, Edward L. Ullman, was pledged to continue the traditional informality of the Young Geographers. Deasy saw to it that as many as possible of the members of this group knew of the differences in the platforms of the two candidates. In June, 1943, McBryde won the post-card balloting by a narrow margin.

Many geographers who were not members of the Association were reluctant to change the order lest they incur disapproval from what they considered the "establishment." Some seemed to believe sincerely that the Association should not be confronted with a rival organization. Consequently, McBryde consulted at first almost exclusively with his close friend and confidant, George F. Deasy, whom he appointed Councillor. McBryde appointed three other councillors, and retained one from the officers of the Young Geographers. John C. Weaver, of the American Geographical Society, secretary-treasurer of the Young Geographers, was continued in that office (as secretary) for the few remaining months of his previous term (though he was too fully occupied at the A.G.S. to contribute more than a mailing list of Young Geographer members). The name "Young Geographers" was replaced by "American Society for Geographical Research" in 1943. McBryde rejected the name "Geographic Research Associates," which the Young Geographers had proposed in May, before the election. He then formulated a resolution and published it in the first issue of the new *Bulletin of the American Society for Geographical Research:*[12]

It is the aim and resolve of the members of the American Society for Geographical Research to acquire and disseminate new knowledge of the earth's surface and its inhabitants, to the limit of their ability, and to encourage in every way possible the younger geographers of the Western Hemisphere. Research in Geography and related disciplines is the unifying bond of the Society, and current geographical investigation or experimentation on a scientific plane shall constitute the only prerequisite for active membership.

The first meeting of this group, named the American Society for Geographical Research, was held on September 16, 1943 in the Cosmos Club Auditorium in Washington, D.C. Nelson H. Darton, one of the original members of the AAG, addressed this first annual meeting. Carl Sauer approved, "I am glad you had old Darton talk to you," then predicted a successful future for the organization. More than 120 persons attended (the Society's first Directory in December 1943 listed 139 charter members; four were inadvertently omitted, making 143 charter members[13]). Of the charter members, 73 were in governmental services (primarily in Washington though most had come recently from colleges and universities), 56 were located in colleges and universities scattered throughout 21 states, and 10 were working in organizations and businesses including the Institute of Pacific Relations, the American Geographical Society, the Tennessee Valley Authority, the Smithsonian Institution, and Safeway Stores. Some distinguished geographers were included among the charter members: George B. Cressey, W. Elmer Ekblaw, Lester E. Klimm, Roderick Peattie, Carl O. Sauer, Helen M. Strong, and Samuel Van Valkenburg. Other charter members who were later to make noted contributions include: Andrew H. Clark, Edward B. Espenshade, George Kish, Shannon McCune, Arthur H. Robinson, and Guido G. Weigend.

Officials for the American Society for Geographical Research invited geographers to membership, stressing its functions as a service organization and a research society. Many geographers felt that definition of research standards would limit membership. Realizing that a distinction should be made between professional and non-professional geographers, it became apparent that a "professional" society was needed. The organization adopted a new constitution and the name was changed to the American Society for Professional Geographers early in 1944.[14] McBryde remained President and E. Willard Miller was Secretary. Board members were Charles F. Brooks (Harvard University), Edward B. Espenshade (Army Map Service), George F. Deasy (War Department), Louis O. Quam (United States Navy), Otis P. Starkey (War Department), and William Van Royen

95

*The American Society for Professional Geographers, 1943-1948, and
The Association of American Geographers, 1944-1948: A Time of Change*

(University of Maryland). Determination, capacity, and growth characterized the new society.

Reaction

Association members employed in Washington, D.C. before the war, and those who had come to Washington between 1941 and 1943, were fully aware of events. The new Society had made itself available to all geographers and that included Association members. The latter feared that there were not enough geographers in the country to support two professional organizations, two sets of publications, two sets of meetings, and that competition might lead to strife. This was the first institutional threat with which the Association had ever been confronted. Otto E. Guthe, then Assistant Chief of the Division of Geography and Cartography, U.S. Department of State, who represented the Association on the National Research Council, (the Association was also represented on the American Council of Learned Societies and the Social Science Research Council) wrote to Association Secretary Ralph H. Brown:[15] "I feel you should be informed of the situation which has arisen regarding the newly established American Society for Geographical Research." Guthe informed Brown that meetings of the Society had been held in Washington. Meanwhile Brown was receiving correspondence from other Association members wondering what was to be done. Stephen B. Jones wrote to Brown:[16]

> . . .looking through the membership roster of the Association, it seems to me obvious that the present election system is no system. There are members who have not published since election. Others have been elected on their second trial, though they were little, if any, changed from the men they were when first rejected. . .I am inclined to plump for the creation of associate membership. . .Associate membership should include, I believe, the right to publish in the Annals.

When Eugene Van Cleef[17] wrote to Brown about the Society requesting advice as to whether Association members should join, Brown replied:[18]

> Perhaps the Council will frame some recommendation. In any event, I hope and pray that the Council will be endowed with more than the usual amount of wisdom in handling this matter. Perhaps this is the spark which will ignite the Association into renewed activity. The new group, by the way, reportedly cannot function altogether successfully as a research group. I wonder if we have not been resting on our laurels too securely?

As one means of meeting the new situation, I am recommending to the Council that the Annals be opened to non-member contributions on a plan similar to the one governing the introduction of papers at the annual meetings.

John K. Wright, Director of the American Geographical Society, after talking with John C. Weaver, a member of his staff and secretary-treasurer of the new Society, wrote to Brown:[19]

I feel that it would be unfortunate for a new and formally organized nation-wide society of professional geographers to be set up alongside of the AAG, and that it would benefit the geographical profession if the two societies could be amalgamated under the name AAG. . .I do feel. . .that it would strengthen the AAG to open its doors to these younger people.

Charles F. Brooks, a long-time member of the Association and a member of the newly created Washington society, wrote to Association President Hugh H. Bennett (with copies of the letter being sent to Robert S. Platt, Ralph H. Brown, Guy-Harold Smith, Vernor C. Finch, T. Griffith Taylor, S. Whittemore Boggs, J. Russell Smith, Robert B. Hall, Edwin J. Foscue, F. Webster McBryde, Preston E. James, George B. Cressey, and Glenn T. Trewartha):[20]

This new Society, it seems to me, is directly the result of the Association's having made insufficient provision for the encouragement of young geographers. I believe this has grown largely out of the overcrowding of the program of meetings, with the result that papers of the youngest members of the profession have been cut off. There has also been a raising of the standards for admission to the Association to a point where membership seems so unattainable as to lead many of the young geographers to feel left out. I have had the unpleasant experience of recommending young, and not-so-young men of attainment and promise, once, twice, perhaps more, only to have them turned down. The waste of effort in providing the details required and writing an adequate recommendation has been great. The discouragement of the candidates has been worse. . .

The young men have seen the need and have taken the matter into their own hands in a laudable manner. If their activities can be included in the Association of American Geographers, as, for example, taking in their budding organization as the Junior Section of the Association, and making the present Association the Senior Section, I think it would be a very satisfactory merger. . .I make this as a definite proposal. . .

Brooks' proposal was not accepted. J. Russell Smith replied to Brooks with copies of his letter to all those to whom Brooks had written:[21]

Thanks, much thanks, for that good letter of September 4th calling attention

97

The American Society for Professional Geographers, 1943-1948, and
The Association of American Geographers, 1944-1948: A Time of Change

to the biting impeachments that have been levelled at the AAG, first by the organization of the young geographers, and lastly and more extremely pointedly by McBryde's proposal.

United we may stand, but divided as we now are almost any one may run away with us, witness the social studies fad, and anyone may neglect us, witness our poor status in the arts college.

Much of this trouble can be traced to the essentially snob-o-cratic theory of our organization and the inertia that has followed.

Thanks for the attempt to wake us up. We must wake up or sleep a deader sleep.

Since none of the Association officers worked or resided in Washington, D.C., an informal group of twelve members of the Association drafted a memorandum entitled, "Principles, Policies, and Criteria on the Election of Members to the Association of American Geographers for Recommendaton to the Council, 24 January 1945." The document was re-drafted by a subcommittee of four members for "Twelve Washington Geographers." Richard Hartshorne functioned as chairman of this subcommittee,[22] which met with others of "the twelve" on several occasions. This document did not recommend a change in principle in the selection of Association members, but did urge reconsideration of the "criteria for measuring work." Prior to this they began a search for geographers who might be eligible for election which uncovered a substantial number of potential candidates. Early in the summer of 1943 the secretary received an unprecedented 53 names, which were submitted to the Credentials Committee. The Committee recommended the nomination of 33 of the 53 proposed geographers. The ballot that was distributed to the membership had never previously included so many names. On November 18, 1943, Richard Hartshorne sent a letter to all those who had sponsored candidates for membership urging them to return the ballots which had been distributed so that they could be counted by the Secretary.[23] He warned that the size of the ballot might encourage an increase of votes for rejection especially among those of the membership unfamiliar with the emergence of a new society in Washington, D.C. (ten percent "no" votes of all votes cast constituted a rejection for membership). Thirty-one of the thirty-three geographers nominated by the Council were elected, yet there were at least half a dozen candidates who would have suffered rejection with just three more "no" votes. That meant Association membership had increased from 167 in 1941 to 176 in 1942, and 207 in 1943. Membership of the Association was very low compared to membership figures of other professional associations. In 1943 the American Historical Association had a membership of 3,615; the American Economics Association, 3,599; the American Sociological Society, 950; and the American Chemical Society, 31,109. The American Astronomical Society had a membership more than three times as large as the Association, though there were fewer astronomers than geographers in the United States—and the prestige of American astronomy remained high.

Two new committees on membership problems were appointed by the Association in 1943 and 1944. In the fall of 1943 a Committee on Membership Planning was appointed, chaired by Edwin J. Foscue, whose task it was to identify geographers who might already have met the qualifications for membership. To facilitate the search the country was divided into seven regions, and a committee member placed in charge of the search in each region. In this way it was thought that no potential members of the Association would be overlooked. Early in 1944 Derwent S. Whittlesey appointed a committee to review the matter of requirements for Association membership. The Committee on the Bases of Membership (chairman Loyal Durand, Jr., Edward A. Ackerman, Carleton P. Barnes, Walter M. Kollmorgen, and Peveril Meigs) offered a preliminary report which was adopted by the Council at its meeting in October 1944, and which was then sent to the membership: "*Report of the Committee on the Bases of Membership*" (October 10, 1944)

    I A Geographer is
        1. a person who has taken at least a Master's degree in some branch of geography; or
        2. one who has had professional employment in some recognized aspect of the field of geography; or
        3. one who has extended geographic knowledge as a result of exploration, writing, or research.

    II Qualifications for Membership
        1. Publication of original research in a recognized journal or book.
           The term publication shall be construed to mean:
          a. a minimum of three research papers, of which not more than one is a direct portion or excerpt of a Master's or PhD thesis, and at least one of which shall

99

*The American Society for Professional Geographers, 1943-1948, and
The Association of American Geographers, 1944-1948: A Time of Change*

          meet the quality standards set below in II 2b. (Further studies along the lines of a thesis are satisfactory.)

  b. publication of a book in which research ideas are processed, and which is not primarily a compilation.

2. Quality of research
    a. The quality must be judged by a credentials committee, or other duly constituted committees, and the Council on the basis of the available publications.
    b. An original work which adds new ideas or new primary data to the knowledge of the profession shall be considered as satisfying part of the requirements whether published in article, journal, or book form.

3. Sponsorship
    A prospective candidate must be sponsored by two members of the Association unless he is in a border field, in which case
    a. he must have three sponsors
    b. the Council must have assurance of his interest in geography.

III Workers engaged in confidential government work
    1. It is assumed that, in normal times, a government-employed geographer will have the same opportunity as all of being appraised as a result of his publications. They are to meet the same standards as others.
    2. In times of national emergency (such as war) workers on confidential reports are eligible upon the following conditions:
        a. they must have published at least one article or book under the conditions set forth above, available to all the profession;
        b. they must be sponsored by a minimum of three persons familiar with their confidential work;
        c. a statement must be presented by the sponsors certifying that:
            (1) the reports contain original work in keeping with the spirit of the Association;
            (2) the reports are not primarily compilations; and
            (3) the reports have been in large part actually written by the candidate.

The Association had made its decision. It would not amalgamate with the new society; it would not offer an amended variety of membership. It had made a scour for eligible geographers, and it had revised the criteria which determined eligibility.

The American Society for Professional Geographers, 1943-1948[24]

The American Society for Geographical Research represented the real beginning of the American Society for Professional Geographers, the new name adopted early in 1944. "It was the formalization of the first utilitarian society for all professional geographers."[25] The development of the American Society for Geographical Research was due in very large measure to the work of F. Webster McBryde, who served as president, secretary and treasurer (one year) and editor of the *ASPG Bulletin* during the formative years (1943-1945). Though McBryde from the outset had envisaged a well-rounded professional service organization, he thought it best to develop it through the intermediate stage of the American Society for Geographical Research, for many geographers were hesitant to join a full-fledged challenge to the established Association of American Geographers. McBryde was employed in the War Department: the chief of his branch was Col. Sidman P. Poole. When McBryde requested of his chief a day's leave to initiate a manpower survey and census of geographers, classified by technical capability, Poole was enthusiastic. McBryde's basic motive was to comb the War Manpower Commission files for potential new ASPG members. "It seemed Poole had been turned down ('blackballed,' he said) three times by the AAG as unqualified." He informed McBryde: "You not only have my blessing for the Society, which can fill a need never met by the A.A.G., but I see in it great opportunities to contribute to the war effort. Take all the time you need. This will be one of your official assignments."[26] The claims of war-oriented research were reduced on McBryde in late spring, 1943, when Allied control of North Africa virtually cut off all threats of German invasions of Latin America, his area of special responsibility. His full-time use of his War Department office for purposes of developing the new geographical society was vital.

In the process of expanding his mailing list for the ASGR at the War Manpower Commission (WMC), National Roster of Scientific and Specialized Personnel, McBryde also recruited for Col. Poole, helping to build the staff of the Topographic Branch; culled the voluminous "geographer" file, at WMC request, and, also at their request, wrote a "Description of the Professional Occupation of Geography." This document became a National Research Council

6. F. Webster McBryde, Founder of the American Society for Professional Geographers (Photograph by Al Schwartz, 1955)

project for inclusion of Geography as a critical wartime profession, and provided the Federal Civil Service with a definition of Geographer for its Register of Professionals. His recruitment program for the War Manpower Commission included the ghost-writing of an article for *Mademoiselle Magazine* on opportunities for women in geography, followed by a 45-minute television program on this subject in New York. At WMC he started a job file and clearing house for employment of geographers, which later became known as the AAG *Jobs in Geography.* The *ASPG Bulletin* which he established and edited evolved into the monthly *Professional Geographer,* as well as the annual Directory. It was McBryde who brought in William Van Royen to assist with the new constitution and by-laws; who set up the seven original regional divisions, recruiting most of the staff of each; and who organized a foreign affiliate organization program which was terminated by the merger. Two of these institutions, in Peru and Ecuador, are still thriving. William Van Royen, after numerous attempts to liberalize and modernize the AAG, was invited by the ASPG president to direct the task of writing a constitution and by-laws for the new society. This he did in conjunction with the officers of the ASPG, mostly during the late summer, 1944. Upon the merger in 1948, the new AAG adopted these instruments with little change. Van Royen's contribution in the development of the ASPG as a professional organization was of major importance. He, George Deasy, later treasurer, and E. Willard Miller, who came in during the second year as secretary, were among the chief organizers of the ASPG.

Growth of the American Society for Professional Geographers was rapid. By November, 1945, membership was 300; by November, 1946, membership was 605; and by November, 1947, membership was 805.[27] When, in December, 1948, amalgamation with the Association took place, the ASPG numbered 1,094 members.[28] Although approximately half of the charter membership of this group was from Washington, this ratio did not continue. By December, 1948 E. Willard Miller calculates that "about 90 percent of the professional geographers in American colleges and universities were members of the society."[29]

The publications of this Society consumed 50 to 70 percent of its total expenditures. The first publication of the Society was a single mimeographed sheet issued in October, 1943. The first major publication of the Society was a 63 page *Geographical Research Directory* in December, 1943, which listed the charter members of the Society and provided brief biographical data for each. During 1944 the *Bulletin* changed from a mimeographed to a printed publication. In 1945 the

Bulletin was expanded to a length of several pages; in November the second *Directory* was issued. In 1946 the *Bulletin* of the Society was transformed into a larger journal with the title, *The Professional Geographer.* Additionally an informal newsletter was inaugurated, and in November 1946 a new *Directory* was printed.

In 1945 a joint meeting was held with the Association at Knoxville, Tennessee, and in 1946 a second meeting was held with the Association at Columbus, Ohio. In 1947 both groups met at Charlottesville, Virginia (the program included a joint meeting with the National Council of Geography Teachers) and in 1948, the year of amalgamation, a joint meeting was held with the Association at Madison, Wisconsin.

In addition to publications and annual meetings, the American Society for Professional Geographers developed regional divisions, which reduced the isolation of geographers. Among the earlier divisions were the Middle Atlantic Division, Northeastern Division, Southeastern Division, East Lakes Division, and the South Central Division. Boundaries of regional divisions were sometimes the object of controversy. From comments made on the membership amalgamation questionnaire, it is clear that the regional divisions were an immediate success.

Committees, too, were part of the Society. These included the Placement Committee, Committee on the Survey of Professional Work (appointed to assist the Research and Development Division of the War Department and the National Research Council), a National Atlas Committee (joint with the AAG), the Government Documents Committee, Committee on Cooperation with the Pan American Institute of Geography and History, Public Relations Committee, Social Science Yearbook Committee, Committee on General Education Reports, and committees on membership, nomination and credentials. E. Willard Miller writes of the Society's committee work: [30] "The committees were responsive to the needs of the time. As I remember one of the most gratifying aspects of being an officer in the Society was the willingness of geographers, not only to accept an appointment to a committee, but to devote great energy and time to its purpose."

The Association of American Geographers, 1944-1948

Those same conditions which had given rise to the American Society for Professional Geographers—rapid expansion of the field of geography, large numbers of geographers not members of the Association of American Geographers, concentration of young geographers in Washington in government work during World War II, concern for professional aspects of geography as well as for scholarship—also had an impact on the activities of the Association. During 1945 the AAG nominating committee, chaired by W.L.G. Joerg, sought a secretary for the AAG who would both represent the traditions of the Association and yet be thoroughly familiar with the recent developments in geography in Washington and elsewhere. The committee chose the young geographer, Chauncy D. Harris, who had taught at four Midwestern universities, and who, by war-time work in the Office of the Geographer of the Department of State and the Office of Strategic Services, had participated in numerous meetings and discussions of the remarkable concentration of geographers in Washington, D.C. during the war years. Harris was proposed by the nominating committee, and elected Secretary of the Association at the meeting in Knoxville, Tennessee, December 1945. At the same time John K. Wright was elected President of the AAG and Stephen B. Jones, a councillor for a three-year term. Wright, Harris, and Jones formed a very congenial working team.

With the backing of the Council of the Association, Harris expanded the activities of the Association.[31] A 65-page _News Letter of the Association of American Geographers_ was issued in July 1946. This contained extensive reports on Association affairs, activities of the research councils, activities of individual members of the Association, work in geographic centers (universities and colleges, government, business, and other institutions), the activities of other geographic societies (including the American Society for Professional Geographers), research tools, and the address list of members of the Association. This _News Letter_ was widely distributed to 1138 geographers in the United States and Canada—not only to members of the Association, but also to members of the ASPG, and to many geographers in the country not members of either organization.[32] For the first time the records and actions of the Association were widely distributed to the profession at large, and not just to members. Also extensive information was provided on non-Association geographical activities. The _News Letter_ carried the notice on the front cover: "Extra copies of this _News Letter_ are available for distribution to geographers who may be interested in its contents. Names of such geographers, together with corrections to or items for the _News Letter,_ should be sent to the Secretary." The _News Letter_ both improved the image of the Association and aided the ASPG in increasing its membership, since through the pages of the _News Letter_ many geographers in colleges and universities first became aware of the existence and activities of a professional geographical organization to which they could belong.

Similar newsletters were issued in the fall of 1946, in June 1947 (listing among other items the volumes in the deposit library of the Association of American Geographers in Cincinnati), and October 1947, providing reports on the activities of regional divisions of the ASPG, recent publications of members of the Association, and an address list of members of the Association. This innovation was highly successful and, on the invitation of the Secretary of the AAG, the ASPG joined as co-sponsor for 1948 of a *Joint Newsletter of the American Society for Professional Geographers and the Association of American Geographers,* edited by Charlotte M. Burtis and Clarence B. Odell. This reported extensively on affairs of the AAG, affairs of the ASPG, joint affairs of the ASPG and the AAG, general news items, centers of geography, news of individuals, news of meetings, and announcements. Four numbers were issued during the year with 135 pages of news. It was a most welcome result of cooperation between the two organizations.

The Secretary of the Association, in order to open sessions of AAG meetings as widely as possible to papers by geographers, personally sponsored the papers of many non-members, particularly of those being considered for membership (such sponsorship was then still necessary for a non-member to present a paper).

At annual meetings of the Association during this period presidential addresses were delivered by John K. Wright (1946), "Terrae Incognitae: the Place of the Imagination in Geography," Charles F. Brooks (1947), "The Climatic Record: Its Content, Limitations, and Geographic Value," and Richard Joel Russell (1948), "Geographical Geomorphology."

At the 1946 AAG meeting in Columbus, Ohio, 67 papers were on the program, 44 by members and 23 by non-members (introduced).[33] Furthermore many of the papers were by younger members who had been in Washington during the war years. Nearly half of the members of the Association were in attendance (119 out of 260).[34] The American Society for Professional Geographers and the National Council of Geography Teachers met at the same time and place. The programs were separate but members of each organization were invited to programs of the other organizations. Total registration for the geography meetings was 480.[35] The Secretary of the Association prepared a mimeographed list of all who registered for any of the meetings and this was widely distributed.

The 1947 meeting in Charlottesville, Virginia, was a joint meeting of the Association of American Geographers, the National Council of Geography Teachers, and the American Society for Professional Geographers, with a common printed program, though some of the sessions were sponsored by only one of the co-operating geographical organizations; others were joint of two organizations; and yet others were joint of all three. The program included 116 papers.[36]

The 1948 meeting in Madison, Wisconsin, was a joint gathering of the Association of American Geographers and the American Society for Professional Geographers. The fully integrated program was arranged by a joint program committee under the chairmanship of George F. Carter. Fewer papers were given, 67, but to facilitate discussion, abstracts were printed in advance in the program.[37]

During this period about a dozen committees were active.[38] Of particular importance were those concerned with membership planning, that is particularly with searching out and identifying geographers who might qualify for membership in the Association (Edwin J. Foscue, chairman, 1943-1946; Stephen S. Visher, chairman, 1947-1948). With the election of 23 new members in December 1945, 14 new members in April 1946, 15 new members in March 1947, and 34 new members in March 1948, the membership of the Association increased from 223 near the end of 1945, to 306 on March 15, 1948.[39] This growth by more than one-third in just over two years was substantial in terms of the past size of the Association, but the Association still included only a small fraction of the active geographers of the United States.

Thus the most important committees of the Association of American Geographers in these years were concerned with the relationship with other geographical organizations, particularly with the American Society for Professional Geographers, and with possible amalgamation.

Toward Amalgamation

After 1943 professional opinion was divided on the question as to whether or not the geographic profession should be represented by the organizations, the Association of American Geographers, and what became the American Society for Professional Geographers. Many Association members felt there was room for only one of the organizations, and most of "McBryde's outfit" were insistent on retention of their organization, else they would belong to none. Feelings ran high. Association members were referred to as "super Ph.D.'s" and "honorary members." The charge was levelled that

107

The American Society for Professional Geographers, 1943-1948, and
The Association of American Geographers, 1944-1948: A Time of Change

many Association members did not attend their annual meetings (and it was a matter of record that Harlan H. Barrows, Isaiah Bowman and Carl O. Sauer had each recently displayed an absence for ten consecutive years). Members of the ASPG were referred to as "Young Turks," "Upstarts" and worse. Unpleasant letters were sent to McBryde, who was also urged by individual Association members to cease his campaign. Clearly, resolution of this problem would not be easy.

Chauncy D. Harris, Secretary of the Association, prepared an informal letter summarizing the situation as it existed in the early weeks of 1946. His position was conciliatory:[40]

It seems to me that the Association has two responsibilities: (1) to encourage high standards of research and scholarship and (2) to provide an organization which effectively represents the entire geographic profession. The Association has emphasized the first to the neglect of the second. In the early days of geography in the United States the high qualifications for membership in the Association served the principal need of the time—the establishment and raising of professional standards and the increase of the prestige of the discipline. The rapid increase in the number of geographers combined with the relative stability of the size of the Association, however, has resulted in the inclusion of a declining proportion of geographers and its failure to establish contact with many members of the profession; notices of its meetings, for example, reach only a small fraction of a body of geographers. . . .

No organization effectively reaches the entire body of the geographic profession. The National Council of Geography Teachers is open to a wide membership but its primary concerns have been with elementary and secondary teaching and an increasing proportion of geographers are finding employment outside the teaching field. The American Geographical Society, open to all, has performed notable work, but it is scarcely an association of professional geographers. The National Geographic is merely popular. The Young Geographers Association filled a real need by its sponsorship of activities for young geographers fresh out of graduate school or yet in advanced study. Its successor, the new Society for Professional Geographers, is full of energy and youthful enthusiasm and will stimulate many good ideas, but only a decadent and neglectful A.A.G. would abandon its rightful role to a new organization, which never would have been conceived or formalized if the Association had discharged effectively its responsibility to represent the field of geography.

Many professional associations have no formal achievement requirement for membership, but in view of the long tradition of the A.A.G., I think it would be a mistake to shift entirely to open membership. Nevertheless some means ought to be found of providing a wider membership base. The Association can carry on its tradition of recognizing scholarship and also become the association of all professional geographers by the creation of two types of membership. Personally I favor the division into fellows and members, as is done in many academies of science.

Negotiations preparatory to amalgamation began seriously in 1946. The five geographers who were of critical importance to the success of this enterprise included Chauncy D. Harris (University of Chicago, and Association member since 1943), Stephen B. Jones (Yale University, and Association member since 1938), John K. Rose (Senior Economic Analyst in the Department of Economic Warfare, Washington, D.C. and Association member since 1943), William Van Royen (University of Maryland and Association member since 1940), and John K. Wright (American Geographical Society and Association member since 1925). It was largely due to the patient and optimistic exchange of ideas among these men that a professional unity was reestablished in geography in the United States. Other geographers, too, were involved, and the process of negotiation was not always easy. E. Willard Miller, Secretary of the ASPG, wrote to Chauncy D. Harris on July 31, 1946, concerning the proposed merger. . ."the full, constructive and friendly cooperation between the two organizations is certainly to be desired." He also stated that the idea of merger had been proposed by the Association and that the Society had little to gain from such a plan. On August 3, 1946, Harris wrote to Miller that "if there is not substantial agreement among members of the Board of the ASPG that amalgamation would promise to serve better than two separate organizations let us not proceed further. . .Will you please let me know at your convenience the decision of the Board whether (1) discussions of possible amalgamation are now terminated, (2) a joint committee should now be appointed to make a more detailed study of amalgamation, or (3) a membership-wide joint poll should be organized. . ."

On September 16, 1946, John K. Wright, then President of the Association, wrote to Chauncy D. Harris expressing his thought concerning the restrictive membership policy:

> The policy of restricting membership has not had any very substantial effect in raising scholarship standards, or if it has done so to some limited extent, the good that has accrued has been outweighed by all the bitterness and cleavage in our profession which has resulted from this policy.

Shortly thereafter, on September 23, 1946, a meeting was held in Washington by a Joint Problems Committee. John K. Wright had selected Chauncy D. Harris and Lester E. Klimm to join him in representing the Association. John K. Rose, President of the American Society for Professional Geographers, was joined by Kenneth J. Bertrand, George F. Deasy, Arch C. Gerlach, E. Willard Miller, James A. Minogue, Louis O. Quam, and William Van Royen, in representing the Society. The purpose of the meeting was to explore and define the problems associated with a possible merger of the two organizations. It was agreed at the meeting that the geographic profession would be better served by one organization than by two, and that a merger of the Association and the Society was desirable. Discussed at length were the different ways in which a compromise could be accomplished concerning the vexed issue of membership policy. Lester E. Klimm (Chairman of the Association Committee on Constitution and Policy) stated that he felt a constitution, acceptable to both organizations could be extracted from the constitution of the ASPG.[41] It was agreed that the presidents of both organizations should appoint members of an AAG—ASPG. Joint Problems Committee which would be scheduled to report at a special joint session at the annual meeting of the two organizations scheduled for December 28-30, 1946, at Columbus, Ohio. Pursuant to that report, the committee was instructed to prepare a questionnaire which would be sent to all members of the two organizations in order to assay professional opinion. A draft questionnaire was prepared and submitted to the Council of the AAG and the Executive Board of the ASPG. Following approval by both organizations, the questionnaire was printed and, in December 1947, copies were sent to all members of both organizations. It was agreed that Burton W. Adkinson would prepare a summary of the results early in 1948. That summary recorded that there was overwhelming support for a merger of the two societies. Of the 457 ballots returned, 397 were for merger, and 60 against merger. Of the ASPG members who were not also members of the Association, 310 voted for merger and 36 voted against it. E. Willard Miller, President of the ASPG, felt that further analysis was necessary in view of the many qualifying remarks on the ASPG questionnaires. This was accomplished by William Van Royen, assisted by Mrs. Battersby, Mr. Booth, Mr. Calhoun, Mrs. Sutherland, and Mrs. Westbrook, and the resultant ten-page analysis was sent to all ASPG members. Determination of the matter was left to a committee, which had been elected jointly by the members of both societies—Arch C. Gerlach, Chauncy D. Harris, G. Donald Hudson, Carl H. Mapes, and Otis P. Starkey. Chauncey D. Harris was named chairman of this crucial United Geographic Organization Committee. The committee of five was "to draw up plans for lines of action and a constitution for an amalgamated organization."

The first draft of the new constitution, which was substantially modelled after the ASPG constitution, was completed by the committee on May 28, 1948. On June 3, 1948, the draft was sent to all the members of the AAG Council and the ASPG Executive Board. On June 15, 1948, the ASPG Executive Board rejected adoption of the constitution because the name "Association of American Geographers" had been adopted. Intense negotiation followed. Arguments advanced for retention of the title of the Association were numerous. Deposits in savings banks and stock certificates were in the Association's name; representation on the National Research Council and on the American Council of Learned Societies was in the Association's name. In academic, governmental, and international circles (for example, the International Geographical Union) it was the Association's title which was known. In libraries, nationally and internationally, it was the *Annals of the Association of American Geographers* which had secured a place, and won many subscriptions. Yet there were many members of the ASPG who were resolutely opposed to retention of the title "Association of American Geographers." It was finally agreed to put the matter of the name of the organization to a vote of the ASPG membership. From a return of 227 ballots, 82 were in favor of "American Association of Professional Geographers," 17 were in favor of "American Society for Professional Geographers," and 128 voted for "Association of American Geographers." Following this vote ASPG President E. Willard Miller, wrote to each of the members of the Executive Board,[42] "I strongly recommend that the Board pass the proposed constitution as it stands by a unanimous vote so that the new amalgamated society can come into existence at the Wisconsin meeting." The board gave its unanimous approval, as did the Council of the AAG.

Officers of the newly constituted "Association of American Geographers" were drawn from the officers elected for 1949 by the two societies:[43]

Office to which elected	*Office in the new Association*
President AAG	President
President ASPG	Vice-President
Vice-President AAG	Councillor for two years
Vice-President ASPG	Councillor for three years
Secretary of ASPG	Secretary
Secretary of AAG	Councillor for one year
Treasurer of AAG	Treasurer
Treasurer of ASPG	Position of Retiring President (i.e., Councillor for one year)
Three elected Councillors of AAG	Elected Councillors
Seven Regional Chairmen of ASPG	Ex-officio Councillors

At the business meeting of the 45th Annual Meeting of the Association, December 29, 1948, at Madison, Wisconsin, the merger of the AAG and the ASPG was unanimously approved. That same evening at the annual banquet, president-elect Richard Hartshorne said of the new Association:[44]

> In urging that our new association should zealously retain all that is best in the heritage of both the former organizations, I suggest also that we may well use the time of change to delete whatever is unproductive in the heritage of each.

> I rejoice that in this new Association of American Geographers we are all fellow members on the same basis. I do not say of equal standing, for that would be absurd. There are those sitting before me whose accomplishments in advancing geography are recognized by all of us as eminent; these men require no labels. Likewise there are those among us who have not yet attained the maturity of thought or who have not yet had sufficient time to achieve equal distinction, but may do so, or more, hereafter. When you have done so, you likewise will need no labels in order to be recognized.

> In particular our constitution makes no distinction—and I as your president will recognize no distinction—between those who yesterday were members of the AAG and those who were not; it follows, I trust, that no one will make a distinction between those who yesterday were members of the ASPG and those who were not. From this time on we all stand on a common footing in one organization. I give you, fellow members, the long life and success of the new Association of American Geographers.

With the passage of time wounds were healed, though there were those who felt that upon merger the "new" Association of American Geographers had sought to expunge the contribution of the ASPG from the record. These individuals offered as evidence adoption of the pre-1948 AAG seal, retention of the 1904 founding date and "founded in Philadelphia," elimination of the names of ASPG officers from post-1948 AAG lists, adoption of the title *The Professional Geographer* (New Series) then later, elimination of the "New Series," and the lack of invitation to past ASPG presidents to sit at the President's Table at post-1948 AAG annual banquets.[45] Importantly a new organization now existed. E. Willard Miller has observed:[46]

> The A.S.P.G. demonstrates that a few dedicated people can make fundamental changes in a profession. A small group of geographers were dedicated to expand the role of the geographic profession and provide new directions that are now taken for granted. But to effect change ideas had to be presented that were accepted by a large percentage of the geographers of that day. The acceptance of the ASPG goals and objectives is evidence that the Society fulfilled a basic need for the furtherance of the geographic profession.

The newly constituted Association combined the selective and conservative elements of the organization which William Morris Davis had founded, and the multi-purpose service organization initiated by F. Webster McBryde. Institutional evolution had wrought a new Association of American Geographers.

7

*The Reconstituted Association of American Geographers: 1949-1963**

The emerging character of the reconstituted Association of American Geographers can be more meaningfully understood if viewed in the context of international and domestic events of the 1950s and early 1960s; the very substantial changes in the nature of geographic thought and research during the 1950s; and the increasing expectations of the steadily growing membership of the Association after 1949.

The 1950s and early 1960s witnessed revolutionary changes in the nature of geography and in geographic research and instruction. First, there was the recognition of the limitations of areal differentiation of the earth's surface as the overriding problem in geographic research, and its consequent impact on the teaching of regional courses in our schools and colleges. In place of the study of the variable character of the earth's surface, there came to be a growing emphasis on the analysis of the spatial dimensions of human behavior and institutions with a concentration on topical geographic problems. Problems involving the distribution of phenomena and the spatial aspects of human and natural processes were studied in the abstract, leading to the development of generalizations, principles, and theories in such areas as spatial diffusion, spatial interaction, and spatial structure. Research focused on the processes involved in the distribution of a given class of phenomena, and the analysis of covariation among processes as reflected in space relations.

Secondly, there was a change in methodology during the 1950s which became a divisive force in the geographic profession. The change involved the greater use of more complex quantitative techniques in modeling causal relations. As pointed out by Garrison,

---

*Chapter by Clyde F. Kohn

"while applications of experimental and statistical methodologies appeared sporadically in the geographic literature prior to 1950, the decade of the fifties was the one in which they were widely adopted and subsequently adapted."[1]

The nature of the Association was further affected by continued growth in its membership and in the non-academic employment of many of its members. At the time of the merger of the Association of American Geographers and the American Society for Professional Geographers, the AAG had a membership of 306; the ASPG of 1094. Because there were many overlapping memberships in the two organizations, the total at the inception of the reconstituted Association is estimated at about 1300 members. At the end of 15 years, this number had more than doubled. The Executive Secretary reported that as of December 31, 1963, there were 2818 members in the AAG. This increase in membership gave rise to management as well as functional problems during the period 1949-1963.

Organizational Problems Resulting From Merger and Growth

As noted in the preceding chapter, Richard Hartshorne, the incoming president of the reconstituted Association of American Geographers, broke with precedent and addressed the membership informally at its annual banquet. Prophetically, he singled out two topics which were soon to become problems. One was related to the nomination and election of officers for the new organization; the other to the definition of membership. Both of these had been debated rather spiritedly during the writing of the constitution for the merged organization.

The Dispute over Nominations and Election of Officers.

Presidents of the AAG, prior to its merger with the ASPG, had been selected by a Nominating Committee appointed by the AAG Council. Single nominations were made for the Association's offices. Those selected for President were primarily distinguished geographers who were expected to present a scholarly presidential address.[2] Ability to direct the affairs of the organization was considered of lesser importance. The chief business of the organization was the responsibility of the Council, but more realistically the Secretary was influential in setting policies and in directing the organization's affairs. As a result of these practices, Hartshorne in 1948 could rightfully say that the AAG "had no candidate for office."

The ASPG, on the other hand, had more than one nomination submitted for its offices, and its Nominating Committee was also selected differently than the AAG's Committee. In the ASPG, a Nominating Committee of three was elected at the business meeting, by secret ballot and by a majority of all Professional Members and Members present, from a slate of five persons or more. Two of these five were proposed by the Executive Committee, and three or more were nominated from the floor. All members of the Nominating Committee had to be Professional Members, and none could be members of the Executive Board. The Committee had to secure written acceptances from the nominees it proposed.

The Constitution of the new AAG provided that the Nominating Committee *should endeavor* to make two or more nominations for each office (Article 4, Section 2). This statement had been debated vigorously by the committee charged to write the new constitution, and represented a compromise on the part of AAG representatives. The Constitution also provided for additional nominations made in writing by any ten members of the new Association.

In light of the first year's experience of the new Association, the Council in January, 1950, appointed a Constitution Review Committee. [3]

One of the problems which surfaced at the October, 1951, meeting of the Executive Committee emanated from a petition, initiated by Richard Hartshorne, chairman of the 1951 Nominating Committee, and signed by 68 members, mostly academics. It called for an amendment to Article 4, Section 2, of the Constitution, requesting that nominating committees in the future need not be under constitutional obligation to find multiple nominees for AAG offices. The petition grew out of the experience of the 1951 Nominating Committee which had found it difficult to secure two or more individuals who would compete against each other for office. The petition was referred to the Constitutional Review Committee.

A second petition, circulated among non-academic geographers, the majority of whom were employed in the Washington, D.C. area, was signed by 168 members and was received by the Council on February 17, 1952. It had been initiated by F. Webster McBryde, William Van Royen, and Hoyt Lemons, and was diametrically opposed to the first. It asked that the Vice-President be nominated as a single candidate for President, but that multiple nominees for the office of Vice-President be made, with a candidate to be selected from a slate of at least three nominees. Those responsible for the second petition thought it unnecessary that the President be an outstanding scholar in the discipline, and suggested that this office might be held by an individual who could manage effectively the affairs of a large and growing organization.

The two petitions were earnestly debated, sometimes even acrimoniously. Sponsors of Petition number 1 believed that the AAG President should be someone recognized as an intellectual leader, and that such individuals were usually not willing to be nominated in competition nor would they "run again" if not elected. Defeated candidates, they pointed out, tended to become unavailable thereafter for offices in which they might well be suitable and needed. Signers of this petition also pointed out that fulfilling the constitutional directive was a burden for the Nominating Committee.

Those who circulated Petition number 2 felt that the reconstituted AAG had become a different organization from its predecessors, much larger and still growing. Thus, they argued, the President's duties should be largely executive, and the individual elected should be freed from the heavy burden of delivering an address which required lengthy scholarship preparation. They further argued that more democratic concepts should be applied in the selection of the Association's officers, including the assumption that decisions by the whole group are better than decisions by individuals or small groups, and that wide participation of members in the affairs of the Association is better than restricted participation. Debate ensued and the spectre of dissolution of the reconstituted Association arose.

The Council attempted to work out a compromise between the sponsors of the two petitions, but, failing to do so, tabled consideration of the problem until after the meeting of the International Geographical Congress which had been scheduled to meet in Washington in August, 1952.[4]

The item, along with nine other recommended constitutional changes, was submitted to the membership in early January, 1953. The membership was asked to vote (1) for Petition number 1, or (2) for Petition number 2, or (3) for the present Constitution. Records show that 130 members voted for Petition number 1; 295 for Petition number 2; and 11 for alternative number 3. Petition number 2 carried, and Article IV, Section 2 (later to become Section 1) was amended accordingly.

Other items relating to the nomination and election of officers of the new organization were also voted on at the same time. These changed the terms of office to make them coincide with the period between annual meetings of the Association; increased the term of office of the Secretary and Treasurer to three years; changed the time of reporting by the Nominating Committee and of mailing of ballots to the membership; required that the Council's nominations for the Nominating Committee be announced to the membership in time for additional nominations by petitions; and prohibited past presidents from serving on the Nominating Committee.

The approved amendment relating to the nomination of officers did not solve the problem for very long. In November, 1956, the Executive Committee once again voted to submit to the membership an amendment to delete the sentence, "The Vice-President is to be selected from a slate of at least three candidates," and to substitute in its stead the statement, "The Nominating Committee shall make two or more nominations for each office, except for that of President." The change from three to two nominees for the office of Vice-President was approved at the annual business meeting on April 3, 1957.

Again, in April, 1962, the Council recognized the difficulties experienced by nominating committees in getting individuals to be candidates, particularly those who had been defeated in previous years. The President was asked to appoint members to assist the chairperson of the Constitution and By-Laws Committee in drafting alternative proposals for consideration by the Council or Executive Committee.[5]

The Office of Honorary President

With the merger in December, 1948, and the greatly enlarged membership of the new Association, the executive duties of the President hampered preparation of a scholarly paper during his term in office. Discussions led to the recommendation of a constitutional amendment providing for the office of Honorary President, who would be given at least one year to prepare a presidential address. The elected President would then be free to carry on executive duties for the organization.

This amendment was related to the debate over the functions of the elected President of the reconstituted AAG, and was one of the ten items voted on in January, 1953. The last elected president to give a presidential address during the 1949-1963 period was J. Russell Whitaker in 1954. The first of the Honorary Presidents to do so was Derwent Whittlesey in the following year.[6] The office of Honorary President was eliminated after 1966 with the appointment of a full-time Executive Officer to manage the affairs of the Association. Thereafter, presidential addresses were once again delivered by elected presidents, the year following their term in office.

The Question of Membership in the New AAG

Following merger of the two organizations, two classes of membership were constitutionally defined: members and associates. Members were individuals whose proof of "mature professional activity in the field of geography" were (1) a graduate degree in geography or equivalent education; or (2) an undergraduate degree in geography and at least two years full-time service as a professional geographer; or (3) an individual who had made significant professional contributions to the field of geography. Associates were persons who were actively interested in the objectives of the Association but who did not meet the qualifications.

One of the ten constitutional amendments submitted to the membership in January, 1953, was intended, interestingly, to stiffen the requirements for full member status in the new Association. It required that those who held an M. A. degree in geography should also possess at least one year's full-time service as a professional geographer, and raised the experience level of those who held undergraduate degrees in geography to *three years'* full-time service as a professional geographer. The membership approved the change.

In September, 1961, the Council requested the Constitution and By-Laws Committee to draft an amendment removing the requirement of experience in addition to a degree in geography for full membership. Subsequently, in April, 1962, the following constitutional amendment was adopted by mail ballot: "Members. The principal criterion of eligibility for membership shall be mature professional activity in the field of geography." The adoption of this amendment opened up membership to the AAG but left a number of definitional problems unresolved.

Qualifications for membership in the reconstituted AAG were determined by a constitutionally established Credentials Committee consisting of three members elected by the membership. During the period 1949-1963, members of this Committee devoted long hours to fulfilling their responsibilities. The removal of criteria for membership in the years following 1963 brought an end, of course, to this committee.

Growth and Composition of Membership

In the three years following the merger of the AAG/ASPG, the new Association grew from about 1300 to nearly 1700 (exclusive of foreign or retired members). Growth continued during the next decade, but at a slower rate due to attrition by resignation, death, or failure to pay annual dues. Thus, in December, 1953, the membership was reported as being 1,627. Of these, 1,521 had mailing addresses in the United States and its territories. About seven-eighths lived east of the 90th meridian, with secondary nuclei in the San Francisco Bay and Los Angeles areas. Washington, D.C. had the largest concentration with 168 members, and within 20 miles of the District of Columbia there were an additional 33 members.

In June, 1957, 1,838 paid members were reported. By the end of 1963, the membership had increased to 2,775. Thus, from 1949 to 1963, the Association more than doubled its membership. [7]

Fortunately, since 1949 the AAG has issued a number of directories. This type of publication was initiated by the Young Geographers Society in 1940, 1941, and 1942 and later by the ASPG, with directories appearing in 1943, 1945, and 1946. The first directory after amalgamation was published in November, 1949, as a *Supplement* to *The Professional Geographer.* [8] Other Directories were published in 1952, 1956, 1961, 1964, 1970, 1974, and 1978.

An analysis of the membership of the Association, based on the 1952 *Directory,* was completed by Louis Peltier. [9] Of the 1,697 members reported in 1952, the occupation of 362 was not given, leaving 1,335 with employment indicated. Using this number, 1,335, as a base, Peltier found that 45.8 percent of the membership was engaged in university teaching, with another 16.2 percent in secondary school teaching or as students. Of the members reporting occupations, 25.7 percent were employed by city, state, and federal governments. Of these, the majority by far, 23.9 percent, were working for the federal government. Peltier also found that the organization was predominantly a society of young men, with more than half of the geographers less than forty years old. Only a quarter of the membership held doctorates.

Based on 1,706 members who returned cards in 1962 (2,004 mailed), it may be concluded that there was an increase in the percentage in academic positions (including students), from 62 to 69 percent. The percentage in college and university teaching and research rose from 45.8 to 50.9 percent. At the same time, the percentage of members employed in government positions declined from 25.7 percent in 1952 to 19.4 percent in 1961. Those self-employed or employed in industry of all kinds also declined, from 11.7 to 9.7 percent. These changes reflect the increased enrollments at the college level during the 1950s and early 1960s, and the growing demand for college and university instructors.

Regional Divisions

Article VI of the Constitution of the reconstituted AAG, adopted in December, 1948, provided for the establishment of regional divisions to promote the objectives of the Association in their respective areas. Upon establishment of a division, a chairman and a secretary-treasurer were to be appointed by the Council. After an initial term (not to exceed two years) all officers were to be elected by the Members and Associates of the Division. All chairmen, however, had to be Members of the Association.

The plan for regional divisions was established during the 1940s by the former ASPG and was continued, following the merger, as one of the essential and more desirable features of the new AAG. The first of the divisions to be formed by the ASPG was the Middle Atlantic Division, established in October, 1944, mainly to serve the large number of geographers then concentrated in Washington, D.C. Other divisions were soon organized, and by the summer of 1945, the Northeastern, Southeastern, Southwest, East Lakes, West Lakes, and Central Divisions were functioning.

During the first year of the new AAG, some Regional Divisions were relatively dormant; others were notably active. Hartshorne noted in his report at the end of the first year after the merger that it was particularly gratifying that geographers in those parts of the country where remoteness made it difficult for members to attend the national meeting, had flourishing regional meetings. In October, 1949, the Executive Committee moved that an informal gathering be planned for the April, 1950, meeting for an exchange of experiences and discussion of problems common to the several divisions.

The 1948 Constitution gave the Council authority to determine the boundaries of divisions, and very early it was called upon to do so. At its 1948 meeting, a New York City and Vicinity sub-group of the Middle Atlantic Division was given the right to organize as a subdivision. At the same meeting the formation of a New York-Ontario Division was approved. The President of the New York State Geographical Association was named Chairman of the New York-Ontario Division (Quebec was later added), and the meetings, programs and activities of the two groups were subsequently coordinated.

During the early years of the new AAG, a number of problems relating to regional divisions were encountered. In March, 1951, during the annual meeting, a luncheon of divisional officers was held. This was chaired by Walter W. Ristow, Secretary of the AAG for 1949-1950, and Chairman of the Middle Atantic Division for the year

1951. Chairmen of divisions, it was noted, served also as *ex officio* members of the Council with the same voting rights as elected Councilors. Walter Ristow discussed the responsibilities of the AAG Council relative to the divisions.

Problems common to the divisions in 1951 included (1) membership in the AAG of divisional members, (2) divisional publications, (3) divisional procedures, (4) divisional boundaries, (5) term of office for divisional officers, and (6) policies regarding the admission or establishment of new divisions. These problems were briefly discussed at the annual business meeting which followed, and a committee was named to consider them. The Vice-President of the AAG was appointed to serve as chairman of this committee.

An early attempt to draw a map of the regional division boundaries was made by Vice-President Loyal Durand, Jr., in 1951 after much correspondence with division chairmen on the areas of overlap and blank spots. A systematic attack on the problem of divisional boundaries was begun in 1953 with a study by Herman O. Friis of the distribution of members of the Association and of the Middle Atlantic Division. Based on the Friis study, a new map of regional divisions was drawn, approved by the Council, and published in the 1956 *Handbook-Directory.* Since 1956, several important additions have been made as Canadian Provinces were admitted into the regional structure of the Association. In 1959, the Executive Committee voted to merge the New York Metropolitan Division and the New York-Ontario Division into a New York-New Jersey Division. Later, this division became a part of the Middle States region. Still later, Ontario was assigned to the East Lakes Division. In 1962, the Province of Manitoba was welcomed into the West Lakes Division, and later the Pacific Coast Division was enlarged to include British Columbia, the Yukon, Hawaii and Alaska. Alberta and Saskatchewan were included in the Great Plains-Rocky Mountain Division.

Central Office Development

Until the merger of the AAG and the ASPG, neither organization had an established national office. The affairs of both were conducted by their respective secretaries and other officers out of their homes or professional offices. However, a need for a central office, and even a permanent secretariat, had been recognized even before the merger of the two organizations. At its meeting on December 20, 1948, the ASPG Executive Committee proposed that the new AAG Council give serious consideration to the establishment of such an office. A Committee on a Central Office and Secretariat was appointed in October, 1949, to explore the problem.

On recommendation of this Committee, chaired by Chauncy D. Harris, the Council voted in January, 1950, to (1) establish a Central Business Office, and (2) establish such an office at Colgate University under the direction of Shannon McCune, then editor of *The Professional Geographer,* Vice-President of the Association and a member of Colgate's faculty. Mrs. Evelyn Petshek was hired to manage the office work and became the first paid employee of the Association.

In late 1950, Professor McCune resigned his position in Colgate University, and an *ad hoc* committee was appointed to investigate the relocation of the Central Mailing Office, as it was then known, in Washington, D.C. The decision was made to establish the office, rent-free, at the Map Division, Library of Congress, under the general direction of the AAG Secretary, Louis O. Quam, and the immediate direction of Arch C. Gerlach, Library of Congress, who had succeeded McCune as editor of *The Professional Geographer.* Mrs. Petshek, who in the meantime had also moved to Washington, D.C., continued as office manager. The need for some consideration of a new Central Office was again raised in the spring of 1955 when Mrs. Petshek resigned. President Louis O. Quam noted in 1956 that the organization was at an awkward size. The work load justified a full-time executive secretary and editor, but income did not permit the necessary action. President Clarence F. Jones echoed the sentiment in 1957 in his address to AAG members, "Our Association is too large to operate without a Central Office, and it is too small to afford a full-time Administrative Secretary."

The Executive Committee at its meeting in November, 1957, appointed a Committee consisting of Wallace W. Atwood, Jr., Arch C. Gerlach, and Chauncy D. Harris, to investigate possibilities for relocating the Central Office, possibly in Washington, D.C., New York, Philadelphia, or in one of the several universities which had offered space and facilities. After careful consideration, the Committee recommended that the Central Office remain in Washington, D.C., and that suitable space be rented for its headquarters. In August, 1958, the Council authorized a subcommittee, consisting of the President, Secretary, and Treasurer, to act for the Association in bringing about the relocation of the Central Office as quickly as possible. This Committee elected to rent space from the National Academy of Sciences in the American Council on Education Building, 1785 Massachusetts Avenue, N.W., Washington, D.C. The Central Office was moved into this rented office space early in 1959.

It was soon recognized that the new office space was not adequate, and in his report to the Executive Committee in September, 1959, President Paul Siple spoke to the need of giving serious thought for additional staff and more adequate space without increasing cost to the members.

At the April, 1960, Council meeting, Meredith F. Burrill, then Secretary of the AAG, reported on personnel, equipment and space problems of the Central Office. President Jan O. M. Broek interjected comments on the need for a paid Executive Secretary even if the costs had to be financed by capital reserves until such time as the office could pay its own way by obtaining research contracts, foundation support, and a larger membership. No action was taken, however, at this meeting.

At the first meeting of the 1961-1962 Council, in September, 1961, under the chairmanship of President Gilbert F. White, a plan to employ a full-time Executive Secretary was discussed. Four steps were planned: (1) to draft a precise description of the position, (2) to solicit comments and suggestions from the membership, (3) to explore sources of funds, and (4) to consider possible candidates. Subsequently, the Executive Committee was authorized to proceed to appoint an Executive Secretary, and to finance a new management program. A list of nine possible candidates for the position was considered.

In January, 1963, President Arch C. Gerlach announced that the Executive Committee had selected Arvin W. Hahn to be the Association's first Executive Officer, and that Dr. Hahn had accepted a one-year appointment beginning on August 5, 1963. In addition to the Executive Office, the office force in November, 1963, consisted of two full-time assistants, Mrs. Betsy Shaw and Mrs. Elizabeth Beetschen, and two part-time assistants. Mrs. Beetschen has continued to serve the Association as its office manager since that date.

The year 1963 was made all the more eventful by the announcement that the National Geographic Society had generously offered the Association rent-free space for the Central Office in its building in Washington, D.C. The move was made early in 1964.

Financial Status and Problems.

Despite expenses incurred resulting from the amalgamation, the report of the Treasurer for the first year of the reconstituted AAG indicates that the organization was financially sound. The annual dues were $7.50, providing an income of $7,579, or 68 percent of the Association's receipts. Another $902 (8 percent) was produced from investments, and $2,610 (24 percent) from the sale of the *Annals*. Including cash on hand and the market value of securities held by the Association, its total value amounted to $32,194.50. A third printing of *The Nature of Geography* had been financed entirely from the 1949 budget, and had within the year returned nearly 45 percent of its original cost.

In 1950, the AAG committed $10,000 to the National Research Council to be used by the U.S. National Committee of the International Geographical Union to arrange for the 17th International Geographical Congress to meet in Washington, D.C. (1952) The Treasurer paid the first $2,000 of this commitment in 1950.

To meet the increasing costs of operating the Association, the Executive Committee increased the dues of Members to $10.00, but left those of Associates at $7.50, effective January 1, 1952.[10] The subscription price to non-Members for the *Annals* was raised to $5.00, that for *The Professional Geographer* to $3.00.

Beginning in 1954, the AAG developed serious financial problems, even though the net worth of the organization, including the market value of securities, continued to increase. There was a drain on cash assets caused primarily by the publication of *American Geography: Inventory and Prospect.* Total receipts for 1954 were $23,888.07, but the disbursements amounted to $30,153.14.

In November, 1956, the Treasurer pointed out that during 1956 and preceding years, the budget of the Association had been balanced by draining reserves, and, according to the auditor's report of 1956, expenditures had exceeded total income by $1,600 for the period 1950-1955. Until 1956, shortages had caused no real problem because the AAG had been able to call upon its reserves in cases of emergency. However, the cost of operating the Association had increased each year from 1950 to 1956, and if additional normal income were not to be found, deficit financing would become the rule rather than the exception. The auditor noted further that in 1956, the Association would have to call on reserves to the amount of nearly $5,000. At this rate, he reminded the membership, the AAG would become insolvent in six years.

President Clarence F. Jones, in May, 1957, had this to say about the financial condition of the organization:

> "Deficit financing in an emergency is necessary and appropriate, but the financial reserves of the Association are not sufficiently large to permit deficit financing to become the rule rather than the rare exception. There are only two ways to solve this problem: one is to increase income; the other is to reduce expenses. A reduction in expenses means fewer services to the members, reduced Association activities, and a smaller publications program. At times

an Association may look to contributions for needed funds. As the Association grows and increases its influence, it may expect more gifts, but gifts normally restricted to specific purposes do not provide cash for what may be called the 'normal operations of the Association.' At the present, our Association must look primarily to its members for financial support."[11]

In order to help solve the problem, dues for members and associates were raised to $15 per year, effective January 1, 1958. By the close of 1958 the organization was once again in sound financial condition and remained so throughout the remaining years of the 1949-1963 era. Membership dues for the year 1958 exceeded $25,500 as compared with $17,400 for the year 1957. Receipts from the sale of the *Annals* and *The Professional Geographer* also showed substantial increases, and donations totaling $4,400 were recorded.

In 1961, the Special Activities fund of the organization rose greatly with the receipt of $55,000 for the High School Geography Project from the Fund for the Advancement of Education of the Ford Foundation.

In 1962, several important gifts and grants were received by the AAG. One of these, for $10,000, was made by the National Geographic Society to support the Association's program to promote geographic education and related research. In addition, grants to support a stronger Central Office and to improve organized geographical research were made by Rand McNally and Company and the A. J. Nystrom Company. Also in 1962, the AAG Council authorized the establishment of a Corporate Membership category with annual dues of $200, and made plans for annual seminars of special interest to business and industry. The first Corporate Member was Holt, Rinehart, and Winston Company, Inc.

In 1963, the AAG received as a second grant from the Ford Fund for the Advancement of Education, $116,000 to finance the second phase of the High School Geography Project. The National Science Foundation in that year granted the AAG $56,660 for a project to upgrade geographic instruction in the liberal arts (see next chapter).

These funds made it possible for the AAG to convert itself from a passive organization into one which became actively involved in promoting geographical research and its application at all levels of education, as well as in business and government. The net worth of the Association as of September 30, 1963, was $147,352.38. By the end of December, 1963, the AAG was indeed in good financial condition. Its total income for that year amounted to $41,039; disbursements to $20,247. This resulted in an excess of income over expenses of

$20,247, making possible the appointment of an executive secretary and two full-time assistants.

Contributions to the Development and Promotion of Geography

Annual Meetings

One of the important objectives of professional and scholarly organizations is to promote discussion among its members and with scholars in related fields. National and regional meetings serve this purpose.

Prior to 1950, the former AAG customarily held its annual meeting in late December and on campuses of major universities. Two major changes occurred in 1950: (1) the annual meeting was held in the spring (April 5-9), and (2) papers presented had to be approved by the Program Committee. They were circulated in advance to selected individuals for the purpose of stimulating and guiding discussion.[12]

The 48th Annual Meeting of the AAG was held jointly with the National Council of Geography Teachers on August 6 and 7, 1952, in Washington, D.C., in the interval between the meetings of the Third Consultation on Geography of the Pan American Institute of Geography and History, and the 17th International Geographical Congress. All sessions were open to visiting geographers from foreign countries.

With the University of Pennsylvania as host, some 550 geographers met in Philadelphia on April 11-15, 1954, to celebrate the 50th Anniversary of the AAG. The Program Committee, under the direction of Andrew H. Clark, arranged a program of invited papers representing "the highest standards of thinking and research in American geography at mid-century." Three general sessions, in addition to 12 group sessions, were held. The general sessions dealt with "Breaking New Ground in American Geography," "Problems in Geography in the United States," and the "Pan American Institute of Geography and History."

McGill University and the University of Montreal were co-hosts to the 52nd Annual Meeting in Montreal, April 1-5, 1956. This was the first AAG meeting to be held outside the United States, and it was also the first for which a registration fee ($2) was charged. In addition to papers presented in general sessions, 62 papers were read in concurrent sectional sessions. Five of these were given in a session entitled, "Statistical Methods in Geography," a session chaired by Harold H.

McCarty, and the first to be held on this topic at an AAG national meeting.

The 54th Annual Meeting in 1958 was a summer meeting, and took place in Santa Monica, California, August 17-22. It was the first meeting of the AAG to be held west of the 92nd meridian. A featured general session, arranged by William L. Thomas on "Man, Time and Space in Southern California," consisted of five papers later published as a *Supplement* to the *Annals.*

The 55th Annual Meeting in 1959, hosted by the University of Pittsburgh, was special in several ways: (1) It was the largest Association meeting held to that date, with some 850 persons in attendance; (2) The program presented the largest number of papers (totaling 149) ever read up to that time at an Association gathering; and (3) No plenary sessions were scheduled apart from the general welcoming session. A joint session was held with the Regional Science Association, featuring a discussion of "Regional Science Techniques Applicable to Geographic Studies."

As a result of the large number of papers presented at the Pittsburgh meeting, the Council created a Committee on Programs to draw up a set of regulations of annual meetings, and to summarize convention activities which had proved most successful or most disappointing during the recent years. Members of this Committee were past chairmen of program committees. This Committee recommended that more time at annual meetings be given to topics of national and international importance; urged that papers of less then outstanding quality not be presented; asked Program Committees to screen papers submitted for the annual meetings in order to assure high quality and reduce the need for concurrent sessions, and sought means for publishing selected papers in *Proceedings* of the annual meeting. Council members, however, were not unanimous in favor of adopting these recommendations. Subsequently, a short article by Preston E. James was published in *The Professional Geographer* entitled, "Remarks on the Presentation of Papers".[13]

In preparation for the 57th Annual Meeting, held in East Lansing, August 28-September 1, 1961, the Program Committee, under the chairmanship of Eugene Cotton Mather, requested that both abstracts and complete texts of papers be submitted not later than April 15, 1961. Undoubtedly, this requirement considerably reduced the number of papers presented. Whereas a total of 148 papers had been presented at the 1960 meeting, only 80 papers were given in the two plenary and 15 concurrent sessions in 1961. It did, however, enhance the discussion and the character of the presentation.

7. The Banquet at the Annual Meeting of the Association, Santa Monica, California, August 1958

An examination of titles of papers read at the annual meetings held from 1949 to 1963 reflects the major trends characterizing the nature of geography during that period. There was a decided decline in the number of papers presented in regional and physical geography, and an increasing number that dealt with the results of theoretical research based on analytical methods using mathematical techniques and models. Increasing attention was given, as for example at the 1962 annual meeting in Miami, to the use of the computer and other technological developments in geographical research. Finally, a growing interest in the role of geography in education, at both the pre-collegiate and collegiate levels can be noted, as well as in the application of geography in business and government. The annual meetings during the 1949-1963 era, therefore, played an increasingly important role in promoting the discussion and sharing of research with colleagues within the discipline.

AAG Journals and Monographs

One of the prime objectives of an organization like the Association of American Geographers is to provide for the publication of scholarly and professional studies pertinent to the interests of its members. It is not surprising, therefore, that one of the first actions of the Council of the reconstituted AAG was to vote to continue the publication of the *Annals,* the principal publication of the former AAG, and *The Professional Geographer,* the official journal of the former ASPG. Henry Kendall of Miami (Ohio) University was named editor of the *Annals;* Shannon McCune of Colgate University of the new series of *The Professional Geographer.* Two advisory Editors, H. F. Raup and Louis O. Quam, were named to assist Kendall in the selection of articles for the *Annals.*

The Annals: In the first year after the merger, several adjustments had to be made by the Editor of the *Annals.* From a circulation of just over 700 in 1948, the number of subscribers jumped to 1700 in 1949. This growth made it impossible for one individual to accomplish, in his spare time, the entire range of duties of an editor which involved serving as critic, reviewer, editor, copy editor, review editor, circulation manager, sales manager, address clerk, mail sorting clerk, shipping clerk, and bookkeeper! Thus, in his first report in 1950, Kendall recommended that an Editorial Board be established with a rotating membership, and that this Board be asked to read manuscripts and make recommendations about their publication. He also strongly urged that a central mailing office be established to take care of routine clerical and business matters.

Based on Kendall's recommendations, the Council in March, 1951, voted to replace the Advisory Editors with an eight-man Editorial Board from which two persons were to retire each year. Manuscripts submitted for publication were to be sent to two members of this Board to be approved or rejected for publication.[14] Clerical and business details were transferred to the newly established Central Office, thereby freeing the Editor to perform his other duties more effectively.

In April, 1954, the Council established the position of Review Editor, naming Norton Ginsburg, University of Chicago, to this position. The first review-article appeared in December, 1956. It dealt with "Current Literature of Communist China," reviewed by J. B. Spencer, University of California at Los Angeles.

A second major innovation of the 1950s was the publication of Map Supplements to the *Annals,* with Erwin Raisz as the first Map Editor. He was succeeded in August, 1958, by Richard Edes Harrison. The first of these Map Supplement Publications accompanied the March 1960 issue of the *Annals* and was entitled, "Cyprus, A Landform Study," by Norman S. W. Thrower. Three such map supplements were published between 1960 and 1963.[15]

Several other significant changes also took place during the 1950s involving reprints, the publication of abstracts, special supplements, and other innovations.[16]

The Professional Geographer   Prior to its merger with the AAG, the ASPG had published five issues of *The Professional Geographer,* an outgrowth of *The Bulletin,* published by the Geographic Research Associates, a predecessor of the A.S.P.G. Both of these publications served as forums for discussion, with emphasis on the professional application of geographic techniques and materials.

Following the merger, a New Series of *The Professional Geographer* was designed to report on Association business, information of interest to its membership, personal items, and news from geographic centers, with four issues a year in 1949, six in 1950, five in 1951, and six each year thereafter.

During the period 1949-1963, several changes were made in the contents of *The Professional Geographer.* The early issues were devoted mostly to personal and organizational information, although a few short articles and notes were included from time to time. Beginning with the March, 1953, issue, a section called "Professional

Notes'' was introduced, containing longer articles dealing principally with the methods, techniques or tools of geography, or with the Association and its activities. Also, during the 1950s, several issues were devoted entirely to papers of interest to geographers specializing in cartography. [17]

The Monograph Series   In June, 1956, the Executive Committee approved the Publication Committee's recommendation to initiate a program of monographs in cooperation with Rand, McNally and Company. That company had agreed to publish and distribute one book a year in the Monograph Series. The company was given the responsibility of designing, pricing, and printing the volumes after due consultation with the editor. The Association agreed to select manuscripts, edit them, and read proof. An editorial board was appointed in 1956 consisting of Derwent S. Whittlesey, Chairman; F. Kenneth Hare, Gilbert F. White, and John K. Wright. Whittlesey was selected as the first editor of the series. [18] The Board early announced that no dissertation submitted in fulfillment of any academic degree would be considered, and that monographs would normally run between 128 and 256 pages.

In the spring of 1959, the first of the Series appeared, entitled *Perspectives on the Nature of Geography* by Richard Hartshorne. This was followed by *On the Margins of the Good Earth* by Donald W. Meinig (1962), and *Offshore Geography of Northwestern Europe* by Lewis M. Alexander (1963).

Special Publications

During the 1950s, the AAG was responsible for the compilation and distribution of several important special publications. *The Nature of Geography,* by Richard Hartshorne, originally published in 1939, remained in great demand in the 1950s, and by the end of the decade, its fifth edition had been printed.

Another special publication, *American Geography: Inventory and Prospect,* edited by Preston E. James and Clarence F. Jones, was published for the Association by Syracuse University Press in 1954, during the 50th anniversary of the AAG. The book had originally been planned for publication in 1952 to commemorate the centennial celebration of the American Geographical Society. Initially, the book had been conceived by James as a publication of field studies to be screened and discussed by a committee. The project was not funded, however, so the idea was abandoned. In its place, an entirely new project emerged, involving competitive discussion but not field work. It was designed to study the objectives and procedures of research in the principal branches of geography, and to define fields for future study. Its development became the responsibility of individuals serving as members of nine NAS/NRC committees, coordinated by James. Expenses for the preparation of the manuscripts were paid from a grant from the NAS/NRC. [19]

During the period 1949-1963, the Association also distributed a brochure, *A Career in Geography.* This was published in 1954, and revised in 1962. Other publications, included *Status and Trends of Geography in the United States,* published in 1959 as Part II, Vol. XI, of *The Professional Geographer,* in collaboration with NAS/NRC, and subsequent reports in 1961 and 1963. During this period, two Handbook-Directories were published, one in 1956 and the other in 1961, both edited by Herman Friis.

Awarding Honors and Research Grants

In accordance with provisions in the new Constitution, the first Council of the reconstituted A'AG, in 1949, established an Honors Committee, with Guy-Harold Smith, Chairman. This Committee was asked to formulate policies in respect to honors and awards. Comments were also solicited from the membership as to the number and kinds of awards to be made.

AAG Honors The Council, in March, 1951, voted to establish an Award for Outstanding Achievement in recognition of individuals who had made contributions of major importance. The first recipient of this award was Gladys M. Wrigley, who in 1952 was honored for her achievements in geographic writing and devoted service to the science of geography. Other recipients of the Award for Outstanding Achievement are listed in Appendix I.

In addition to the Outstanding Achievement Award, the Association in 1952 initiated a program to recognize those who had made outstanding contributions toward the progress of the profession by citing them for Meritorious Achievements. The first to receive Citations were Edward A. Ackerman, Lloyd D. Black, George F. Jenks, and Clara E. LeGear. A list of those who subsequently received Citations for Meritorious Achievements appears in Appendix I.

Research Grants During the early years of the new AAG, few requests for assistance were received by the Committee on Research

Grants. This lack of applicants was undoubtedly due to the fact that, at best, grants could support only a minor research project, or assist in only a small way with a larger one. Also, at this time, considerable sums for research were being made available through government and other channels.

In 1956, two gifts were received. Mrs. W.W. Atwood, Sr., gave $1,744 to increase the principal of the Atwood Research Fund to $5,000 and to provide a $500 award in 1957. Also in 1956, the A.J. Nystrom Company Fellowship Fund of $2,400 was established to promote the study of geography in the secondary schools. In 1957 the Council agreed to establish a scholarship in mathematical and statistical geography, and referred the action to the Fellowship Committee. There is no record, however, that the AAG ever made such an award.

The Work of Special Interest Committees

Like most scholarly organizations, the Association has contributed to the advancement of geography through the work of special interest committees: three technical committees warrant mention.

One of the first of the new committees to be established after the merger of the AAG and ASPG was a Committee on Cartography, formed as the result of a petition by a group of some 30 members concerned with the development of cartography and geography. The Committee was given the responsibilities of (1) stimulating interest in cartographic research and publication, (2) assisting in the development of a section on cartography at the annual meetings of the AAG, and (3) cooperating with the Executive Committee of the Division of Cartography of the American Congress on Surveying and Mapping. During the first half of the 1950s, the Committee devoted much time and effort to its many assignments, and in 1957 on its recommendation a Map Editor was appointed to assist the Editor of the *Annals* in the development of a Map Supplement.

The National Atlas Committee

Two years before the AAG/ASPG merger, the two organizations established a joint National Atlas Committee, chaired by S. Whittemore Boggs, to make preliminary plans for a *National Atlas of the Geography of the United States,* comparable to such other national atlases as the *Atlas de France.* At first, the work of this committee was constrained by its inability to secure funds even to prepare an initial dummy. Fortunately, in June, 1948, the American Council of Learned

Societies provided $2,500 to help prepare a prospectus and complete an illustration. The plans that were developed called for about 470 maps.

Federal agencies were surveyed in 1949 to determine the type of data that each might be expected to furnish. A conference sponsored by the Conference Board of the American Council of Learned Societies, the American Council on Education, the National Research Council, and the Social Science Research Council, was attended by representatives of 28 federal bureaus and agencies. Additional funds ($10,000) were sought through the Conference Board, but that Board concluded that so large a project could be produced only by the government, and that they should not, therefore, attempt to raise the necessary funds. The Atlas Committee, however, continued to seek private sponsorship for a number of reasons: (1) in order to assure independence of scholarship and judgment; (2) to insure a unified presentation of maps, and (3) to guarantee freedom from having to depend on annual government appropriations at a time of rising defense expenditures due to the Korean War.

The original Committee was discharged in April, 1950, and a smaller committee was formed in its place. On the basis of its recommendations, the 1951 Council moved that the National Atlas should be produced as soon as possible, and recommended that the AAG and the American Geographical Society cooperate in developing a plan for its production, and in soliciting funds to support the project. The Director of the AGS strongly supported the project, and urged the Society to take charge of its development and publication. The AAG Committee recognized the leadership role of the AGS, and stated its willingness to cooperate in whatever way possible. In 1952, the AAG Committee was discharged, and a liaison committee was appointed to work with the AGS with Carleton P. Barnes, Chairman.

The AGS approached several large corporations and foundations for support, but by the end of 1952 had not been able to raise the necessary funds. As a result, the AAG investigated the possibility of producing a loose-leaf atlas, the contents of which might be prepared by various mapping agencies in the federal government. Such a publication would make it necessary, however, for the government agencies to publish their maps in a standard format and scale. Responses to inquiries about this were encouraging enough to merit further work on the part of the Committee. John Reed, of the U.S. Geological Survey was asked to work with some of the agencies in order to achieve a common format. Unfortunately, the AGS was not enthusiastic about the development of a looseleaf atlas, and so withdrew from the project.

In April, 1954, the AAG Committee on the National Atlas was discharged when the National Research Council became interested in the National Atlas project and appointed a committee to work on it. Early in 1955, the NRC Committee recommended standards for maps suitable for inclusion in a loose-leaf atlas.

In the spring of 1956, the AAG Council reactivated its Committee on the National Atlas, with Arthur H. Robinson as its chairman. Its purpose was to assist the NRC Committee in recommending desirable maps as to contents, compilers, and printers. Within two years a portion of the Atlas was completed, at which time some 60 sheets were published in a loose-leaf binder, all prepared to Atlas specifications. Forty-one of the sheets were based on the 1954 Census of Agriculture. Other maps were contributed by the U.S. Weather Bureau, the Department of Agriculture, and the U.S. Geological Survey. Unfortunately, not all of the maps could be obtained from any one place.

Between 1958 and 1961, the NRC Committee on the National Atlas continued under the able chairmanship of Carleton P. Barnes, and by 1961, some 80 map sheets had been compiled. By that time, it became apparent that there was need for a more formal arrangement of the contents, for a more uniform quality, and for centralized distribution. Consequently, the NRC Committee proposed its own termination. The President of the National Academy of Sciences recommended in 1962 that responsibility for the National Atlas be transferred to the United States Geological Survey. That recommendation was accepted by the Secretary of the Interior, and the National Atlas Project was established in the Topographic Division of the U.S. Geological Survey with funds appropriated by Congress. Arrangements were made with the Library of Congress for the services of Arch C. Gerlach, Chief, Map Division, to be detailed to the Geological Survey for approximately two years to serve as Chief of the National Atlas Project. Thus, by the end of 1963, production of a U.S. National Atlas was truly on its way.[20]

Committee on Mass Data Processing Techniques

The Committee on Mass Data Processing Techniques was an outgrowth of an earlier Committee on Punchcard Techniques, appointed in 1958 and chaired by Howard G. Roepke. In 1958, this Committee pursued two lines of inquiry: (1) a study of the kinds of geographic problems whose solutions might be expedited by the use of calculating and computing machines; and (2) a survey of projects already being undertaken by geographers or on geographic subjects using punchcard methods. In 1959, the Committee, whose name had by then been changed, sponsored a panel discussion on "Experience with Electronic Computers in Geographic Research" at the annual meeting in Pittsburgh. The participants in this discussion, led by William L. Garrison, were Brian J.L. Berry, William T. Fay, Edwin N. Thomas, and William Warntz.

In 1960, the Committee conducted a Council-supported survey of research projects in which geographers had used mechanical aids either to deal with large quantities of data or to perform complex calculations. The objectives of the Committee were threefold: (1) to determine the extent to which geographers were using mass data processing techniques; (2) to determine availability of computer programs and of information already prepared for processing, and (3) to stimulate other geographers to use data processing techniques by making them more aware of the range of problems to which these techniques might be applied. Results of the survey appeared in the March, 1961 issue of *The Professional Geographer.*[21] The Committee was discharged in September, 1961, following publication of the results of its survey.[22]

Contacts with Other Scholarly Organizations and Government Agencies

After merger, the reconstituted AAG maintained many of its former contacts with other scholarly organizations. Among these was its association with Section E of the American Association for the Advancement of Science,[23] the American Council of Learned Societies, the National Academy of Sciences and the National Research Council,[24] and the National Science Foundation. Representation on the NRC made it possible for the AAG to advance the role of geography in ways which would have been impossible without its presence. The Academy and Research Council do not automatically do anything for a science, but they are consulted by government and scientific bodies, and by individuals. The NRC makes grants for conferences of its working committees, and can be of great help in getting research projects launched.

During the years 1949-1963, the AAG maintained close and significant working relations with a number of governmental agencies. Limited space makes possible reference to the work of only three of these: the Air Force ROTC program; the Bureau of the Census; and the Command and General Staff School at Fort Leavenworth, Kansas.

A Committee on Air Force ROTC was appointed in 1952 with Jan O. M. Broek as Chairman, to deal with an Air Force course in Political Geography, planned for senior level students in the Air Force ROTC program. The Committee also was asked to consider a geography unit of 10 contact hours required of freshmen in a curriculum which was to go into effect during 1953-1954. The Committee was highly successful in implementing all of its objectives, including its recommendations to improve the quality of the ROTC teaching personnel. As a result of its work, geography courses in the Air Force program were greatly improved; but even more important, many college students who otherwise might have taken no college geography courses had an opportunity to study geography via the Air Force ROTC program.

In November, 1957, a Committee on Military Geography was appointed to provide advice requested by the Command and General Staff School at Fort Leavenworth, Kansas, on a course in Military Geography. Based on the contacts of committee members with the needs of the Staff School, two articles on the potential and nature of Military Geography were subsequently published in *The Professional Geographer.* [25]

An Advisory Committee to the Bureau of the Census was organized in August, 1958. It set up five subcommittees to carry out its activities: (1) a subcommittee on the codification of geographic coordinates, with Brian Berry as chairman; (2) a subcommittee for a census of tourism and recreation, with James R. Wray as chairman; (3) a subcommittee for review of cartographic activities, with Lawrence Wolf, chairman; (4) a subcommittee to provide for the dissemination of Census documents to schools and public libraries, and (5) a subcommittee to arrange a special session for the Denver (1963) annual meeting. A lengthy report of the Geographic Coding Subcommittee appeared in the July, 1964, issue of *The Professional Geographer.* [26]

Relations with Foreign Geographers and Foreign Geographic Organizations

During the 1950s, the AAG encouraged its membership to increase participation in foreign area research, and to establish more contacts with foreign geographers and geographical research. To achieve these objectives, a joint committee of the AAG/ASPG had been established even before merger. This committee was especially interested in helping foreign geographers and institutions of higher learning rebuild their libraries and map collections destroyed or damaged during World War II. Several reports were published in the early issues of the New Series of *The Professional Geographer* relating to foreign geographers, the development of geography in overseas countries, and impressions and recollections of American geographers who traveled abroad.

The Committee on International Fellowship and Research Grants. In 1952, the AAG Council appointed a Committee on International Fellowship and Research Grants to replace an earlier Fulbright Act Committee. Under Donald Patton's direction, the Committee, in 1954, prepared a *Guide to International Fellowships and Research Funds,* listing available opportunities for study, research, and teaching abroad. [27] Supplementary lists were published through 1962 except for three years (1959, 1960, 1961). Donald J. Patton and Walter M. Deshler compiled the lists before 1959; Alexander Melamid authored the one in 1962. The Committee was discharged in April, 1962.

Committee on UNESCO Relations. The Executive Committee in 1949 appointed a Committee on UNESCO Relations to keep the AAG membership informed of activities of that organization, and to make recommendations to the Council for possible ways of helping UNESCO achieve its objectives. J. Warren Nystrom was appointed to serve as Chairman of the Committee. During the 1950s, the AAG sent delegates to the several national conferences of the U.S. National Commission on UNESCO.

In 1953, the Council voted to discharge the Committee on UNESCO Relations, and named Wallace W. Atwood, Jr. to serve as liaison officer to keep the Association informed of UNESCO activities. In 1960, in response to an invitation from the U.S. National Commission, the AAG named a delegate, Clyde F. Kohn, to serve on the Commission. He served for two terms, 1960-1966, and was succeeded by Norton S. Ginsburg, who served for two additional terms.

PAIGH. The Pan American Institute of Geography and History (PAIGH) is another international organization in which the AAG was actively involved during the 1950s. [28] In 1951, the AAG Council adopted a resolution requesting the Congress of the United States to increase appropriations for the Pan American Institute of Geography and History from $10,000 to $50,000.

PAIGH in 1952 recommended that a series of volumes on the Geography of the Americas be developed. In August of that year the AAG Council appointed a committee with John H. Garland as chairman to carry out a suggestion that the United States participate in the preparation of one of the volumes, *The Geography of the United States.* Unfortunately, the AAG lacked the finances to support the programs or regional research that the committee had proposed. In November, 1957, the committee was discharged.

Special Regional Committees. Special regional committees were formed from time to time, during the 1950s, as members expressed a desire to become better acquainted with geographers or the geography of specific parts of the world. Examples of such committees included the Committee on the Near East, the Committee on Asian Studies, and the Committee on Chinese Geographical Literature. Another committee was appointed to develop relations with geographers in the Soviet Union. Attention is here focused on the activities of that committee.

There was already a Committee on Soviet Studies, chaired by Chauncy D. Harris, in existence during the first year of the new AAG. Harris had published a short article in *The Professional Geographer* in January, 1948, on "Geography in the Soviet Union."[29]

On recommendation of the Committee on Soviet Studies, the AAG Executive Committee in November, 1957, expressed its willingness to sponsor an exchange of American and Soviet geographers. In March, 1959, Harris obtained Council approval to request the National Science Foundation for approximately $18,000 to send American geographers to the USSR to study the status and application of various aspects of geography in the Soviet Union, in exchange for a delegation of Soviet geographers to make a comparable study in the United States. The NSF approved the grant. A group of six Soviet geographers visited the United States from June 27 to July 23, 1961. Chauncy D. Harris, representing the AAG, accompanied them throughout their trip. The six geographers included I. P. Gerasimov (Chairman), K. A. Salishchev, F. F. Davitaia, G. A. Mavlyanov, V. A. Krotov, and V. P. Kovalevski.[30]

The American group visited the Soviet Union from August 17 to September 13, 1961. It consisted of Chauncy D. Harris, Chairman, Edward B. Espenshade, Jr., Preston E. James, William Horbaly, Joseph A. Russell, and Harold H. McCarty.

The International Geographical Union. During the 1950s, the AAG was instrumental in strengthening the role of American geographers in the activities of the International Geographical Union. In 1949 the AAG made grants for part of the travel expenses of George B. Cressey and Chauncy D. Harris to serve as delegates to the 16th International Geographical Congress in Lisbon, Portugal, April 8-15, 1949. Cressey was elected President of the International Geographical Union 1949-1952 and past president 1952-1956. Harris later served the Union as Vice-President 1956-1964 and as Secretary-General and Treasurer 1968-1976.[31]

In 1948, the National Academy of Sciences, through its U.S. National Committee on the IGU, issued an invitation to the IGU to convene its 17th International Geographical Congress, scheduled for 1952, in the United States. Letters supporting the invitation were issued by various societies, including the Association of American Geographers. The invitation was accepted by the General Assembly of the Union at its meeting in Lisbon in April 1949. Wallace W. Atwood, Jr. estimated that a budget of $50,000 would be needed to prepare and conduct the meeting. The AAG was asked for financial assistance, and in 1949 authorized a contribution of $10,000 to be made available over a period of three years.

The Seventeenth International Geographical Congress and the Eighth General Assembly of the International Geographical Union were held in Washington, D.C., August 8-15, 1952. Several other organizations scheduled meetings to coincide with the Congress: (1) The Third Consultation of the Commission of the Pan American Institute of Geography and History met from July 25 to August 4; (2) the Centennial Program of the American Geographical Society was held in New York City from August 4-6; (3) the National Council of Geography Teachers held a joint meeting with the Association of American Geographers in Washington, D.C. on August 6-7; and (4) the Seventh International Congress of Photogrammetrists met in Washington and Dayton, Ohio from September 4-16. A large number of special committees of the AAG were responsible for excursions, exhibits, finances, foreign delegations, local arrangements, and publications. In all, more than 125 geographers worked on Congress plans and arrangements. Walter W. Ristow obtained leave from the Library of Congress to supervise the planning and scheduling of the meeting, and a total of 779 Americans participated in Congress activities.

Promotion of Geography in Government and Business

During the 1950s and early 1960s, the AAG maintained an intense interest in promoting geography through four hardworking and productive committees: the Committee on Geographers in Government, the Committee on Geographers in Business, the Committee on Careers, and the Committee on Status and Trends of Geography in the United States.

The Committee on Geographers in Government grew out of an earlier committee on Geographers in National Defense. Arch C. Gerlach, who chaired the Committee in 1956, presented a lengthy report in which he discussed the employment of geographers in the Department of Defense, the Army Map Service, the Office of the Assistant Chief of Staff for Intelligence, the Aeronautical Chart and Information Center, the Air Weather Services, the Military Service, the Library of Congress, the Department of Commerce, the Department of the Interior, the State Department, Foreign Service, and those agencies that had summer and temporary projects.

The Committee on Geographers in Business was appointed in 1956 with John W. Reith, Chairman. The Committee conducted a survey of some 450 members whose professional activities were related in one way or another to business. Of this number, 156 replied, and they formed the basis of a report published by the Association.[32]

Two sets of special publications of the Association during the 1949-1963 era testify to the interest that the organization maintained in promoting geography in government service and business, as well as in education. These resulted from the efforts of the Committee on Careers in Geography and the Committee on the Status and Trends in Geography.

The Committee on Careers had its origin in 1948 as a committee of the National Research Council. In November, 1949, it became a joint NRC/AAG Committee. In trying to implement its objectives, the Committee soon found that there was a lack of basic information, and secured private funds to carry out the fact-finding stage. Subsequently, a selected bibliography on careers in geography was published,[33] to be followed by a 35-page manuscript entitled *A Career in Geography,* May, 1954. This was widely distributed by the AAG, and the Committee was discharged.

In November, 1957, the AAG and the NCGT appointed a joint committee to review the pamphlet, and after several years of study, the Council implemented the Committee's recommendation in 1962, to have the pamphlet revised and printed by the Denoyer-Geppert Company.

The second set of publications dealing with geography and geographers in government service, business, and education was prepared by the Committee on Status and Trends of Geography in the United States. This Committee was asked to prepare a report, at least every two years, for submission through the State Department to the Commission on Geography of the Pan American Institute of Geography and History, and for distribution by the AAG. The Committee's work was directed by Clarence F. Jones. The first of the series dealt with *Status and Trends of Geography in the United States, 1952-1957;* the second with *1957-1960;* the third with *1960-1962.* The last was completed under the direction of John P. Augelli. Financial assistance in the preparation of the reports was given by the National Academy of Sciences/National Research Council. The Advisory Committee in Geography to the PAIGH offered much help in preparing the plan and scope of the reports. The first report appeared as Part II of Volume XI, No. 1 of *The Professional Geographer* for January, 1959; the second was published separately in March, 1961; the third was also published as a separate in 1963.

The three issues of *Status and Trends of Geography in the United States* proved very useful to the membership of the AAG, and especially to graduate students in seeking career opportunities.

The Placement Committee. One of the outstanding functions of the ASPG before its merger with the AAG grew out of its concern about the placing of geographers in both academic and non-academic positions. Hence, as early as 1945, it created a Placement Committee, "to serve as an employment clearing house for its members, and as a central source of information for employers of geographers." It maintained an active file of personnel record cards, and submitted names from it to prospective employers. After reorganization, the new AAG Council continued the Placement Committee, with James A. Brammel as its first chairman.

For the past thirty years, members of the AAG Placement Committee have spent many long hours serving the AAG membership. During the early 1950s the Committee performed an "employment" function. It constructed a Placement Data Card File which incorporated an objective key selector system enabling the Committee to compile lists of candidates for positions on the basis of stated requirements. These lists of qualified candidates were sent by the Committee to potential employers; later the Committee notified candidates of job opportunities and had them assume the responsibility for contacting the employers.

Throughout the 1950s, the Committee urged Departments of Geography to make more effective use of the Placement Committee in meeting their needs for faculty, and in placing their graduate students. Many of the major universities, however, were loath to do so, believing that it was less than respectable for a highly-rated department to make use of such a service, an attitude that was to persist until the 1970s when affirmative action programs forced departments to advertise available positions more widely.

In August, 1958, the Placement Committee, with the approval of the Council, began to circulate lists of positions on a monthly basis, making it possible to inform members of positions as soon as they became available.[34] Those members maintaining up-to-date Placement Data File cards received these lists.

As time went on, the Placement Committee replaced its "placement" function with "job development" activities. Its members devoted much time to locating job opportunities outside the academic field, particularly for geographers with the B.A. and M.A. degrees, and a listing of Federal Civil Service examinations for which geographers were eligible became a regular feature of *Jobs in Geography*. Resulting from efforts, listings in JIG increased from 67 in 1959, to 82 in 1960; 86 in 1961; 112 in 1962; and 144 in 1963. The circulation grew from 230 in December, 1958, to 332 in March, 1961, and to 498 in December, 1963, with issues sent to those who requested it, employers who announced in it, and to about 150 departments of geography. In time, JIG became the standard tool for job recruitment and placement, and it has proved invaluable to all those seeking positions in geography. Today, all of the major universities announce available positions in the publication.

The Promotion of Geography in Education.

Prior to their merger, both the ASPG and the AAG professed an interest in geographic education as well as in research and its application in government and business. The Constitution of the merged organizations continued this commitment by listing among its several objectives, "the application of geographic findings in education."

In keeping with this objective, one of the early actions of the Executive Committee of the new AAG was to send a letter to Earl McGrath, then newly appointed Commission of Education, U.S. Office of Education, expressing the wish that a suitable appointment be made as soon as possible to the position of Specialist in Geography in Higher Education, vacated by Otis W. Freeman in August, 1948. At the same meeting, President Hartshorne was instructed to write the presidents of Yale University, Rutgers, and the University of Delaware commending them on the establishment of Departments of Geography in their respective institutions. Also, coincidentally, the first report of a committee to be published in the New Series of *The Professional Geographer* was that of the AAG Committee on Graduate Appointments, recommending that all departments of geography making such appointments should conform to the action of the Association of American Universities adopted at their 48th Annual Conference in 1947.[35]

President Hartshorne, in his address at the annual banquet in 1948, reminded the members of the AAG that as long ago as 1904 the AAG had been interested in developing better conditions for the study of geography in schools, colleges, and universities.

At the Collegiate and Graduate Levels. Throughout the 1950s, the AAG expressed much interest in the preparation of geographers for research and for positions in teaching, government service, and business. At least three presidential addresses during the period 1949-1963 were devoted entirely or in large measure to graduate training for careers in geography: Hudson in 1951; Sauer in 1956; and Thornthwaite in 1961.[36] A number of papers on undergraduate and graduate instruction in geography also appeared in both the *Annals* and *The Professional Geographer,* and in 1957, an information discussion session was held during the annual meeting on the topic, "Teaching Introductory Geography on the University Level." It was noted that in many universities, undergraduate courses were designed more for majors and graduate students than for non-majors in the liberal arts. *The Professional Geographer* for March, 1958, carried two important articles: "Geography and the Liberal Arts College," by Homer Aschmann, and "Introductory Graduate Work for Geographers," by Gilbert F. White.

A major development in the AAG's involvement in undergraduate education came in 1961 with the appointment of a Committee on Liberal Education, chaired by John F. Lounsbury. This committee was created to explore the proper role of geographic thought and skills in liberal education. To carry out its assignment, the Committee, with the approval of the AAG, in 1963 obtained a grant of $56,550 from the National Science Foundation. A six-day national conference was set for December, 1963, bringing together college and university professors of geography from various geographic regions to identify and discuss specific problems, and to develop guidelines for the improvement and modernization of instructional programs. Special papers were prepared for this conference by William L. Garrison, Arch C. Gerlach, Chauncy D. Harris, Arthur H. Robinson, Gilbert F. White, and M. Gordon Wolman. The project was to become highly successful in later years.

At the Pre-Collegiate Level. The 1950s also witnessed a growing concern in the AAG with the quantity and quality of geographic instruction at the pre-collegiate level. A significant resolution was passed, for example, at the Business Meeting of the AAG in March, 1953: "Whereas

> Geography is not generally taught in high schools in the United States, a country now occupying a position of world leadership in which a knowledge of how humankind has occupied and apportioned the globe and utilized its resources is of paramount importance;
>
> A more widely diversified and balanced curriculum for the future leaders of the nation in business and government, and for good citizenship in general, requires an acquaintance with the geography of our own country and of the world of which we are a part;
>
> Young men and women who serve in our armed forces should have a better understanding of the position of our country and its capabilities, and should have some idea of what ways of living to expect in other countries when assigned overseas;
>
> Our nation does not provide the majority of its young men and women with the necessary knowledge of the geography of the United States and that of other countries and of the world as a whole;

Therefore,

> Be it resolved that the Association of American Geographers further encourages the Department of Health, Education, and Welfare of the United States and the appropriate non-governmental agencies to call this serious gap in our education to the attention of the National Education Association, state educational associations, school boards, parents, and educators and to facilitate remedial action."

During the 1950s, the AAAS, an organization with which the AAG was affiliated, became more and more concerned about the status of science and mathematics in the high schools of our nation, especially in 1958, after the Soviet Union's success in launching *Sputnik.* It recommended to institutions and accrediting agencies that added emphasis be placed on science and mathematics teaching, including the preparation of secondary teachers. It also proposed legislation promoting science education. Alfred H. Meyer, the 1958 AAG delegate to the AAAS, directed the following questions to the AAG Council and Membership: "In view of the recent extraordinary accent on science and science teaching preparation and certification, what new role should the field of geography play (a) in the college curriculum, and (b) in the area of secondary teacher preparation and certification? What action, if any, should be taken by the Council in this matter?"

In 1959, a panel discussion was scheduled for the Annual Meeting of the AAG to discuss the issue, "Teaching Geography in our Secondary Schools." The panel consisted of Preston E. James, Clarence W. Olmstead, and Gilbert F. White, and was chaired by Clyde F. Kohn. White suggested that an action type program be designed leading (1) to the preparation of teaching materials, study guides, visual aids, and other matters which would aid public school teachers in their development of basic geographic concepts and understandings; and (2) to encourage the development of an honors course in high schools in cooperation with local institutions of higher learning which could be used for admission with advanced credit to universities and colleges.

White's suggestions were immediately brought to the attention of the AAG Council by Kohn, who proposed that the AAG appoint three representatives to a Joint Committee to work with three representatives of the NCGE on problems related to the improvement of geography instruction in the secondary schools, and to implement White's two recommendations. The Council approved the establishment of the Joint Committee, and charged its representatives to work out details to be submitted as proposals to appropriate agencies for financial assistance. Kohn, representing the AAG, and White the NCGE, were named as co-chairmen of this Joint Committee.

The Joint Committee, in 1960, was authorized to proceed with negotiations for support of its proposal to develop a high school geography course using TV tapes and associated study guides, maps and globes, and other laboratory materials. White and Kohn presented the proposal to the Fund for the Advancement of Education of the Ford Foundation, and in April, 1961, received a grant of $50,000 to be administered by the AAG, in support of Stage 1 of the project.[37]

Stages 1 and 2 were carried out in 1961 and 1962. In 1963, the Ford Foundation granted an additional $116,000 to be augmented by a third and final grant of $7,000 in 1964, to support the further development of Stage 2 and the completion of Stage 3. On the basis of a Conference (Stage 2) including a number of professional educators, school administrators, high school teachers, and professional geographers, an *Advisory Paper* was prepared by John Fraser Hart and William D. Pattison. The ten experimenting teachers and their academic advisors who had been selected to complete Stage 3, met in August, 1962, to study the Advisory Paper, and to discuss, in detail, the basic geographic concepts that had been presented by Brian J.L. Berry, William L. Garrison, Richard Hartshorne, Duane F. Marble, and Edwin N. Thomas, and to plan their units to be developed in Stage 3.

During 1963, William D. Pattison developed a *Response to the Advisory Paper.* A conference of the experimenting teachers, their advisors, and members of the Joint Committee met in August, 1963, to discuss the *Response Paper* and to state and analyze their reactions to Stage 3. Also in 1963 the AAG Executive Committee, in cooperation with the NCGE, appointed a Steering Committee for Stages 3 and 4, and named Nicholas Helburn to direct the project. Named to the Steering Committee were John E. Borchert, William L. Garrison, Chauncy D. Harris, Preston E. James, Clyde F. Kohn, Fred B. Kniffen, Robert B. McNee, Alexander Melamid, William D. Pattison, Clyde P. Patton, and Gilbert F. White, Chairman.

As a result of these many activities, the AAG became deeply involved during the years 1949-1963 in the promotion of education at pre-collegiate, undergraduate, and graduate levels.

8

New Directions, 1963-1978*

The period from 1949 to 1963 had witnessed an unprecedented expansion of geography in the United States. Undergraduate enrollments increased, graduate departments strengthened and thrived. The new academic directions produced scholars and scientists who engaged in growing numbers in funded research. Whereas hundreds of geographers had formerly been supported by teaching assistantships during their graduate studies, many now received on-the-job graduate training as research assistants in funded projects. More geographers found employment beyond the academic departments that had been the traditional marketplace: in business, in government agencies, as consultants and in numerous other capacities they entered the non-academic world. The "new geography," as some (yet again) described the product of the scientific revolution in the discipline during the late 1950s and 1960s, yielded a modern generation of geographers trained in contemporary methods of theory construction and data analysis.

Diversification and proliferation were the hallmarks of geography during the 1950s and 1960s. From remote sensing and computer cartography to urban planning and location analysis, geography now encompassed new as well as traditional areas. Colleges established geography departments where none previously existed, and the number of doctorates awarded rose markedly during the late 1960s. The growth and complexity of geography put pressure on the Association's Central Office, which, in 1962 was still housed in three basement rooms of the National Education Association's building. As early as August, 1961, the Association's Council had

> . . .Agreed that a full-time executive should be employed; that he should. . .
> expand on a full time basis the present part time activities of the Secretary; that

*Chapter by Harm J. de Blij.

there should be presented at the Association's Business Meeting, for general discussion, the Council's view that those activities should be extended on a broad scale to include the promotion of the Association and the profession vis-a-vis other disciplines and internationally. . .[1]

This proposal led to debate in the profession. There were those who argued that the Association had proved itself able to cope with the new demands and who saw no need to modify its structure. The appointment of a full-time Executive Officer might bankrupt the Association. Others, more optimistic and confident that the era of growth would continue, wanted to seize the opportunity through what they viewed as an investment in the future: a stronger Central Office, more effective representation, greater cohesion, a more effective focus for the discipline. In 1962 the discussion was resolved in favor of the optimists, and in January, 1963, it was announced that the Association's Executive Committee had appointed Arvin W. Hahn as Executive Officer for a period of one year, commencing on August 5, 1963.[2]

Arvin W. Hahn came to the Central Office on a one-year leave from Concordia Teachers College, where he was Chairman of the Division of Social Sciences and Director of Graduate Studies. The Association's finances were strengthened by gifts and grants from various sources, including Rand McNally & Co., A.J. Nystrom & Co., the National Geographic Society, and the United States Steel Foundation, amounting to nearly $20,000 in the year the first Executive Officer was appointed. At the same time, membership for the first time in the Association's history exceeded 2,000: on September 4, 1963, President Gerlach (who was in office 17 months from the April, 1962 Miami Beach Annual Meeting until September, 1963 at Denver) was able to report that the AAG had 2,419 members.[3] In part this reflected the continued strengthening of the Association's regional Divisions. President Gerlach in his 1963 Report identified the West Lakes Division as among the most active in a very successful membership drive. The regional Divisions played a major role in the growth of the Association during the 1960s. Some of the activities of the Association are now treated in chronological order.

Project Development

Several important developments occurred during the Presidency of Arthur H. Robinson and the Hahn administration of 1963-64. The Association accepted an offer of free space made by the National Geographic Society, and plans were made to move the Central Office to the Society's Washington headquarters. The Central Office was to remain there until its present facilities were occupied in May, 1971.

At the same time the National Science Foundation made a substantial grant to the AAG in support of the work of its Liberal Education Committee, chaired by John F. Lounsbury. This was the genesis of the Commission on College Geography, and from its beginnings on June 1, 1963, until its termination in mid-1974, this project received a total of over $825,000 in grants. Simultaneously, Gilbert F. White and Clyde F. Kohn developed a proposal for the improvement of geography training at the high school level. The High School Geography Project became one of the Association's most substantial ventures, aided by National Science Foundation grants totalling over 2.6 million dollars.

The Commission on College Geography (CCG)

The aim of the Commission on College Geography was to improve geographic programs at the undergraduate level. To this end the CCG developed and published three series of publications: the General Series, the Resource Papers, and the Technical Papers. These were designed for widespread use by instructors of college geography and related fields. The papers facilitated the incorporation of recent developments in the field and the results of recent research into undergraduate programs of instruction. The Commission also sponsored several Summer Institutes for College Teachers of Geography, supported by the National Science Foundation and the Office of Education, and organized a series of locally hosted Regional Conferences, designed to improve communication among geographers in large universities, state and private colleges, and junior and community colleges. In addition, the Commission developed a Consulting Service through which consultants provided help and information to geography departments that were expanding their programs or attempting to maintain existing programs in the face of decreasing enrollments and retrenchment. The CCG terminated on June 30, 1974, after more than a decade of productivity, but certain of its programs were continued beyond that date. The Consulting Service was sustained under the auspices of the Association, and additional Resource Papers were published in successive years after 1974.[4]

The High School Geography Project (HSGP)

The question of geography's place in the high school curriculum has always concerned American geographers. During the late 1950s the National Science Foundation sponsored a number of school curriculum reform projects of national scope, and geographers wanted to participate in this significant effort. As revealed in the previous chapter a grant (1961) from the Ford Foundation's Fund for the Advancement of Education facilitated the development of a proposal for a new program in the teaching of geography for the high schools of the nation.[5] The National Science Foundation funded the AAG's project proposal in 1963, and the High School Geography Project continued without interruption until 1970. It was directed by William D. Pattison from 1961 to 1964, by Nicholas F. Helburn from 1964 to 1969, and by Dana G. Kurfman from 1969 to 1970. Gilbert F. White served as chairman of the Steering Committee from 1961 to 1970. Conferences, workshops, and institutes were held in many parts of the country to disseminate the products of the HSGP's authors, and, as in the case of the CCG, the project outlasted its funding period. In 1974, work was begun on the revision of the successful HSGP course entitled *Geography in an Urban Age* with the aid of a substantial advance from the original publishers, the Macmillan Company, under the direction of Fredric A. Ritter.

A number of other AAG-sponsored projects were funded during the 1960s and 1970s, but the CCG and HSGP projects had seminal roles in the development of the Association as well as the profession. Initially the projects' headquarters were based at the home institutions of the principal investigators (CCG at Eastern Michigan University and HSGP at the University of Colorado), but later the Association's Central Office became the administrative focus for AAG projects. The Central Office took responsibility for their fiscal policies and practices and coordinated their academic progress.

During the first year of its operations under a full-time executive officer the Association reaped several benefits, not only in the areas of project development but also from a major membership drive launched by Arvin Hahn. Under the motto of "A Thousand More by Sixty-Four" the Central Office wrote to non-member geographers, to department chairpersons, to potential corporate members, and others who might join the Association.[6] By the end of 1963 this campaign had secured an additional 399 members, bringing the total AAG membership to 2,818.[7] And the drive fell just barely short of its goal: the Association had more than 3,000 members in mid-1964. In addition, the newly strengthened Central Office undertook to promote vigorously the distribution of the *Annals* and *Professional Geographer*. The Executive Officer's report to the president of

February 12, 1964, reflects the rising level of activity in the Association's headquarters.[8]

But increased membership totals and journal subscriptions were not the sole objectives of the Association's central administration:

> . . .Possibly the most significant activity carried on recently has been a series of visits with heads of various sections of the National Science Foundation. These visits were designed to acquaint responsible representatives with the Association and its program and to discuss the possible areas in which geographers may make contacts of value.[9]

Arvin Hahn's successor as AAG Executive Officer, Saul B. Cohen, was nominated during the presidency of Edward B. Espenshade, Jr. Cohen who accepted a one-year appointment in mid-1964, vigorously pursued numerous grant opportunities in various Washington funding agencies and developed a network of contacts that were to benefit the profession and the Association for years to come. The AAG Council encouraged this work, but not at the expense of other services to members. For example, the Council instructed the Central Office to facilitate the preparation and distribution of a booklet entitled *Careers in Geography,* a far-sighted proposal that was implemented by Preston E. James.[10]

The Association's progress continued to be aided strongly by the goodwill of the National Geographic Society. As Executive Officer Saul Cohen reported in July, 1965, the AAG occupied some 1,800 square feet of rent-free space in the NGS headquarters, with adequate facilities to hold committee meetings as well as work and storage space. Thus the Association could build a sounder financial base, and upon advice of the finance committee the Executive Committee recommended that,

> up to half of the Association's capital assets be in the common securities market, of which about one half would be in "blue chip" securities and the other half in "growth" or "high risk" securities.[11]

A renewal grant from the National Science Foundation for the further development of the High School Geography Project made possible a series of trials of the available materials in high schools in various parts of the country. During 1965 and 1966 approximately 3,000 high school students used units on settlements, manufacturing, mining, agriculture, culture, habitat, water resources, and political geography. Evaluations were conducted by the Educational Testing Service of Princeton, New Jersey. The renewal grant also made possible the production of new learning units and the involvement of a greater number of geographers.

An important and productive initiative of the Association also bore fruit in 1965, when the National Science Foundation funded a proposal to organize a Visiting Geographical Scientist Program. This project, which continued until 1972, was designed to enable prominent geographers to visit colleges where geography was not part of the teaching curriculum or where geography departments were comparatively small. During the typical visit, a geographer would, in a one-week period, visit a cluster of three or four colleges and present classroom lectures, seminars, and public addresses. The host institutions arranged informal conversations with students, faculty, and administrators; the net effect was to generate greater awareness of the discipline and to plant ideas that might lead to the further development of geography.

During the existence of the Visiting Geographical Scientist Program, geographers visited 237 institutions to lecture and lead seminars. The impact of the program went well beyond what the original proposal had envisaged, because visitors often expanded their itinerary to visit nearby established departments of geography, there to meet colleagues and discuss mutual professional concerns. The VGSP, thus, contributed importantly to the dissemination of ideas in the discipline in general, and to the diffusion of the results of geographical research to other fields as well.

John Fraser Hart succeeded Saul B. Cohen as the Association's third Executive Officer on September 1, 1965. Membershp approached 4,000, gifts and grants were at unprecedented levels, and work loads in the Central Office were greater than ever.[12] The first meeting of the Association's Executive Committee for 1965-66 addressed the problems associated with success: the need for a larger clerical staff, the potential need for an Administrative Assistant to the Executive Officer, the need for IBM automation for the preparation of a new *Handbook—Directory* of the Association. Some consequential organizational changes also came under discussion, including a revision of membership categories (deletion of corporate and associate memberships in favor of contributing membership) and the elimination of the "President's Program" at annual meetings. Henceforth, the immediate Past President would present a banquet address, releasing the ever-busier President from the responsibility of addressing the membership present.[13]

A Long-Term Executive Officer

Another question addressed by the 1965-66 Executive Committee involved the Association's need for a more permanent Executive Officer. John Fraser Hart was the third incumbent in three years, and although the Central Office had been well-managed by each of the Executive Officers, discontinuity had obvious disadvantages—in the Central Office itself as well as in Washington and on the national professional scene. In May, 1966, it was announced that the search for a long-term Executive Officer had resulted in the appointment, effective August 1, of J. Warren Nystrom as the Association's chief administrative officer.

J. Warren Nystrom was to remain at the helm of the AAG until his announced retirement in 1979. Before his 1966 appointment he had been on the geography faculty at Rhode Island College, Chairperson for ten years at the University of Pittsburgh, Manager of the International Relations Department of the United States Chamber of Commerce and supervisor of eight other departments, and member of the senior staff of the Brookings Institution in Washington, D.C. [14] The title of the position J. Warren Nystrom now assumed was changed from Executive Officer of the Association to Executive Secretary, and its level of responsibility considerably escalated. The AAG Council, as well as its Executive Committee, which had repeatedly expressed concern over the fiscal management of the Association's funded projects, charged the incoming Executive Secretary with the task of establishing appropriate standards and practices applicable to these activities.

When the Nystrom administration commenced, the Association, either alone or cooperatively with other professional organizations, was involved in five funded projects: the High School Geography Project, the Commission on College Geography, the Visiting Geographical Scientist Program, a joint AAG-NCGE writing project funded by the National Geographic Society, and the Consortium of Professional Associations for the Advancement of Teacher Education (CONPASS). The AAG took a leading role in CONPASS, funded by the United States Office of Education in 1966. CONPASS involved a five-association project for the assessment of institute programs and teacher preparation. When it was funded, the AAG became the administrative focus for CONPASS and the new Executive Secretary, J. Warren Nystrom, was co-director on a one-quarter time basis. [15] But the relationships of other on-going projects to the Central Office were far less specific. Budgets were being exceeded, daily remuneration for consultants was inconsistent, and liaison with the National Science Foundation was inadequate. The Executive Secretary saw project management and fiscal responsibility as his first and most urgent objectives, and he developed policies and procedures that would meet the highest standards of independent audit.

The Newsletter

Upon the recommendation of J. Warren Nystrom, the AAG *Newsletter* was revived to become one of the most successful ventures to be launched by the Central Office. From a publication of four-pages in November, 1967, it grew into a monthly publication (bi-monthly in the summer) containing, in the late 1970s, twenty pages at times. When, in 1977, the Council's Long Range Planning Committee sent the membership a questionnaire asking respondents to rank the Association's professional publications, the *Newsletter* placed a clear first—ahead of *The Professional Geographer* and the *Annals.*

The success of the *Newsletter* resulted from the attention given it by Salvatore J. Natoli, who joined the Central Office staff in August, 1969. Natoli, who came to the Association from the United States Office of Education, had been on the faculties of Mansfield State College, Clark University, and the University of Connecticut. His position in the Central Office, as Educational Affairs Director, involved the coordination of on-going education projects of the AAG as well as the editorship of the *Newsletter.* In time he converted the *Newsletter* into one of the best publications of its kind. Commencing with the November, 1971, issue, the publication *Jobs in Geography* (JIG), became part of the *Newsletter,* enhancing its appeal to the membership. In addition, the *Newsletter* carried numerous regular features including Council and Executive Committee Meeting Minutes, News from Geographic Centers, Of Note (individual accomplishments), Grant Opportunities, Forthcoming Professional Meetings, Books Received, Publications Available, as well as news from the Regional Divisions, information on travel opportunities, items from and about other professional journals, and a host of other material. The *Newsletter* has always kept AAG members informed about the work of Association committees, the progress of funded projects, and plans for the forthcoming Annual Meetings. As a vehicle for the diffusion of information throughout the profession, the *Newsletter* has no peer.

Remote Sensing

The year 1967 also marked the approval and funding of an AAG-sponsored project in remote sensing by the United States Geological Survey and NASA. The Project was initiated by Arch C. Gerlach and J. Warren Nystrom. The Commission on Geographic Applications of Remote Sensing, chaired by James R. Anderson, focussed on the development of a land use classification system to accommodate spacecraft-generated data. The director of the project, Robert W. Peplies, announced that a land use map of the United States would be produced following the creation of the appropriate classification system.[16] The project had numerous additional results, because it facilitated the presentation of several workshops designed to acquaint professional geographers with the technology and research opportunities of the remote sensing field. And after the Commission had completed the project's program, in 1972, a permanent AAG Committee on Remote Sensing continued to function, producing several useful publications.

But the Remote Sensing Project was not alone in securing support. The National Science Foundation made two awards during 1967 to the High School Geography Project: in June, $58,000 to support teacher education activities and the preparation of instructional materials, and in October, more than $480,000 to support the entire program. With the appointment of Donald J. Patton as editor of HSGP publications, the AAG entered into a negotiation with the Macmillan Company for the publication of the project's Settlement Theme Course, entitled *Geography in an Urban Age.*

The Commission on College Geography added two important publications to its growing list of published materials for use in undergraduate teaching: *New Approaches in Introductory College Geography Courses and Introductory Geography: Viewpoints and Themes.*[17] In addition, the Commission prepared several course outlines and resource papers that were to become part of a still-continuing series.

Guide to Graduate Departments

The Association's publications included not only its major journals, but also a growing number of items produced and disseminated by the staff of the Central Office. The late 1960s witnessed the production of the first of several significant AAG publications. In 1968 J. Warren Nystrom met with all graduate department chairmen present at the Association's annual meeting. He proposed an annual publication detailing matters related to graduate departments of geography to be produced on a self-supporting basis. This was approved by the departmental chairmen present and by the Council two days later. The first annual *Guide to Graduate Departments of Geography in the United States and Canada* was published in 1968. And in 1969 (see below) the Association's annual *Proceedings* made its initial appearance.

The *Guide to Graduate Departments of Geography in the United States and Canada* was an instant success. Until its appearance, no comprehensive listing of information about graduate geography programs in the United States and Canada had been available to potential graduate students. In September 1968 the Central Office sent questionnaires to 165 graduate departments, and in December the first annual edition of the *Guide* appeared, carrying details concerning individual programs and research facilities, academic requirements, financial aid, and faculty. Because the *Guide* contained information on geographers' areas of specialization and Ph.D. origin, it soon became perhaps the most frequently-used quick-reference volume the Central Office has published, the *Handbook-Directory* notwithstanding. The *Guide* has continued to appear annually, and is made available to members at a reduced cost.

The purpose of the *Proceedings* was to improve discussions at the annual meetings by prior distribution of papers to all Association members and to quicken scientific communications in the field. For some time before 1969 the Council had been considering the desirability of publishing a volume that would contain the papers to be presented and discussed at annual meetings of the Association. These ideas coincided with a study of communications processes during professional meetings, prepared by members of the Johns Hopkins University Center for Research in Scientific Communication.[18] The availability of such a volume would, it was believed, enhance professional communication (surely a primary objective of annual meetings) and render discussion sessions more productive. A three year grant was obtained from the National Science Foundation in 1968 which permitted publication and distribution of *Proceedings* to all Association members. The first issue of the *Proceedings* was distributed to members prior to the 65th annual meeting, in August, 1969. Edited by John Fraser Hart, Volume 1 of the *Proceedings* contained 37 papers and consisted of 168 pages on the format of the Association's *Annals.*

Each succeeding volume of the *Proceedings* was edited by a member of the Program Committee for the Annual Meeting where the papers would be presented, and as the attendance and participation at Annual Meetings grew, the *Proceedings* expanded. The 1975 volume (No. 7), edited by John A. Jakle, consisted of 311 pages and included no less than 63 papers. But it was evident that the *Proceedings* was not achieving what the Council had hoped they would. Time constraints prevented the publication's distribution to meeting participants before the Annual Meetings; with difficulty it was included in the registration package. This was too late for thoughtful reading and the preparation of responses, so that the objectives Council had had in mind were not being met. Furthermore, NSF support for the *Proceedings* ended with Volume 3 (1971), and the production of the *Proceedings* from the Annual Meeting budget proved problematic. By 1974 the Council was debating the future of the *Proceedings;* in April, 1975 the Chairman of the Publications Committee, Richard E. Lonsdale, reported that the Committee had divided 6 to 6 on the question of the experiment's continuation. [19] Council declined, at that time, to take action—and so a 1976 volume of the *Proceedings* did appear, edited by Michael P. Conzen, for the New York City Annual Meeting. The Council took the problem up again during its April, 1976, meeting in New York, and on the narrowest of votes a fateful decision was reached: the *Proceedings* were abolished and hence none was prepared for the 1977 Salt Lake City Annual Meeting. [20] Nevertheless, the *Proceedings'* eight volumes bear witness to the quality and diversity of the Annual Meetings of the late 1960s and the early 1970s.

Geography and Afro-America

If the experiment with an Annual Meeting *Proceedings* was one of the Association's innovations of the late 1960s that failed, another effort initiated during this period had permanent and noteworthy results. For some time a number of geographers had been concerned over the status of the discipline in predominantly black colleges and the very limited representation of black professional geographers on university faculties and elsewhere. During the Washington (1968) Annual Meeting a special session entitled "The Status of Negroes in Geography" focussed on such topics as "The Geographic Dimensions of Race Relations" (addressed by Richard L. Morrill), "Recruiting Black American Geographers" (by Theodore R. Speigner), and "The Geographer in the Community" (by William Bunge). Also at this session, Donald R. Deskins, Jr. distributed *Geographic Activities at Predominantly Negro Colleges* and *Geographical Literature on the American Negro, 1949—1968: a Bibliography.* An early workshop on these topics was held at Clark University, led by Saul Cohen. These activities foreshadowed the development of a project proposal to improve geographic education at predominantly black colleges, and in

1969 the U.S. Office of Education made an initial grant of over $250,000 to the AAG in support of these aims. This grant gave rise to the Commission on Geography and Afro-America, which was well directed by Donald R. Deskins, Jr.

Funded support for COMGA continued until 1975 and amounted to nearly $1.2 million. During its highly successful tenure COMGA offered graduate fellowships at major geography departments to qualified students interested in teaching geography at predominantly black colleges, established a Research Panel that encouraged the development of research topics of importance to the black community, sponsored Leadership Conferences designed to familiarize selected faculty from black institutions with modern thinking and current programs in geography, sponsored a Faculty Exchange Program to enrich existing geography programs at predominantly black colleges, and organized Summer Workshops to inform secondary school teachers of innovative teaching methods recently developed in geography. COMGA also established an Information Service that included a placement service, conducted research on the status of black geographers, assisted in the development of new geography programs, and promoted and disseminated research on black America. The Director repeatedly reported in *The Professional Geographer* and in other geographic publications on the results of COMGA's efforts. Among the positive aspects of COMGA was the decision, by many participating geography departments, to provide matching grants to further support COMGA fellows. In some years, this departmental support totalled some $150,000.

Activism, Ideology, and Division

The 1960s witnessed a gradual but increasingly dramatic change in the AAG membership's perceptions of the role of a scholarly or professional society. The Association's pursuit and acceptance of project grants, for example, produced an interventionist behavior in educational matters. This is not to say that there was no concern for education prior to this period, but the projects themselves were, in effect, an expression of activism applied to AAG objectives rather than passive "advancement" (in the terminology of official AAG objectives). In a sense the language of AAG objectives might be interpreted to read "The purpose of the AAG is to advance professional studies in geography and to encourage the application of geographic research in education, government, and business (by sponsoring national projects to improve the quality of geographic education in the schools and colleges)."

An important outgrowth of the AAG's funded projects was that they involved far more people in the Association's affairs than had ever been the case, uncovered and employed a large array of talented people, and changed the Association from a comparatively closed to an open society. In the process, "professional" interventionism and activism soon extended into other areas as well.[21] The late 1960s were turbulent as well as productive years in the history of the Association. Inevitably the Vietnam War divided the profession as it fragmented every other sector of society in the United States, and annual business meetings began to hear motions and petitions for an AAG position not only on the War, but on other issues as well. A noteworthy case in point was the "Resolution on National Priorities" read by Robert W. Kates during the 1969 business meeting, demanding a reversal of national goals and objectives and approved by the members present, 197 to 62.[22]

The Council confronted a difficult decision following the 1968 Democratic National Convention in Chicago, where violence had erupted and law enforcement appeared to have gone beyond civil norms. The Association was among numerous organizations whose Annual Meetings were scheduled for 1969 in Chicago, and several professional associations had already made known their decision to relocate their conferences as a matter of protest. Officers of the AAG received mail from the membership demanding a similar move, but on September 19, 1968, the Executive Committee declined to take action pending a decision by full Council.

The November, 1968, issue of the *AAG Newsletter* carried a letter from President John R. Borchert reporting Council's reaction: "The Council divided 10—10 on the question of moving the meetings to an alternative location. . .(but) it is the Executive Committee's judgment that a decision to meet in Chicago at this time would result in serious diversion of energies better directed into social geographic research and teaching. . .therefore, the Executive Committee has taken unanimous action to remove the 1969 meetings from Chicago and to accept an invitation from the Department of Geography at the University of Michigan to meet in Ann Arbor, August 10 to 14, 1969.[23]

As President Borchert indicated, the membership was deeply divided on this particular issue as well as others. Many members had opposed a move from Chicago, and nearly 800 voted against the Kates Resolution when it was submitted by mail as a referendum to the membership.[24] Officers, Councillors, and members at large voiced concern over the potential politicization of the Association. In July, 1969, *The Professional Geographer* carried a proposal by President

Borchert "concerning policy and procedure for committing the Association on public issues" that reflects this concern quite strongly: "The Council of the Association normally will not issue statements or take actions which mobilize or commit the Association on a public issue. The exception to this rule will be instances when public issues threaten to adversely affect the profession."[25] The procedure involved, *inter alia,* the requirement that no less than 250 signatures must appear on a petition requesting Association action.

These and other barriers notwithstanding, matters of social and political concern did find their way onto the floor of AAG business meetings throughout the 1970s, ranging from the exclusion of positions at South African universities from *Jobs in Geography* to the condemnation of the Pinochet regime in Chile. Thus the annual business meeting in Boston on April 18, 1971 erupted into angry debate over a resolution. ". . .it is the sense of the Geographers assembled at this 67th Annual Business Meeting of the AAG to inform all other fellow members of the Association that we condemn the perpetuation of American military involvement in South East Asia (and) demand immediate withdrawal of all military manpower, armaments, or economic support of any military activities in South East Asia." A motion of inappropriateness was defeated and the resolution was approved by a vote of 78 yes, 30 no, and 13 abstentions.[26] Nor did the end of the conflict in Vietnam signal the end of ideological conflict at annual business meetings. At the 1975 annual business meeting in Milwaukee, for example, a resolution proposed that the AAG create a committee to examine the relationship between intelligence agencies and geographical science "for the purpose of clarification of the legitimate and illegitimate roles which may occur in such associations."[27] (The motion was approved by those present, 107 yes to 78 no, but was defeated in a mail referendum, 346 against to 229 for). The membership proved itself deeply divided, a legacy of the 1960s that was to afflict the Association for years to come.

Annual Meetings: Dilemmas Aplenty

Ideological differences emerged also in the AAG Council, but on other grounds. Although Council did authorize the formation of a Committee on Marxist Geography and underwent an occasional discourse on the evils of imperialism, it faced urgent and immediate problems—and serious disagreement—in quite another arena. For years, Program Committees had been screening the papers submitted for presentation at annual meetings. As a result, the number of papers accepted for presentation could be limited and the annual meeting held to manageable dimensions. Then, during its 1973 meeting, the AAG Council decided to experiment with an "open convention policy," directing the 1974 Program Committee to accept all papers that conformed with its stated guidelines.[28] The result of this decision was an enormous expansion of the 1974 and succeeding annual meeting programs. In New York City in 1976, for example, there were

40 volunteered paper sessions with 262 papers

32 special sessions with 95 papers

5 poster sessions with 95 papers

18 special interest group sessions

9 committee business meetings

6 workshops

2 general sessions (plenary and business)[29]

The 108-page Program of the 1978 Annual Meeting in New Orleans listed well over 1400 participants, and attendance records were set repeatedly during the 1970s.

But all this growth had its price. The Association's Council—and the membership at large—remained divided on the question of the open convention. Many felt that the routine acceptance of all submittals had led to a decline of standards; others asserted that an Annual Meeting Program Committee should organize and facilitate, not restrict and reject. In late 1975, President Marvin W. Mikesell in a letter to the membership described the alternatives and appealed for an expression of opinions:

> . . .Proponents have felt that a program open to all comers. . .is a healthy situation for the Association. A place on the program might be regarded as a legitimate claim, particularly of younger AAG members, for whom the right to be heard might mean the right to be hired. . .(But) a program restricted to a smaller number of presentations would have the advantage of being less troublesome to those responsible for program planning and local arrangements. . .Brilliance and ineptitude, excitement and boredom, standing room only and nearly empty rooms—these have been the obvious contrasts of our recent meetings. But what is the purpose of our meetings?. . .[30]

What, indeed? There could be no doubt that growth in the 1970s also meant change. Twenty-two concurrent sessions at the 1978 New Orleans annual meeting included such topics as "Research in Latin America by Women," "Radical Perspectives in Geography," and "Gay Geographers;" poster sessions provided authors an opportunity to display rather than read their papers; technical workshops in remote sensing and other areas permitted interested geographers to become better acquainted with modern hardware and data analysis. So enormous had the range and variety of geographic interests become that no generalization could suffice. In the absence of a control mechanism, geographers brought any and all material to the annual meetings, and a kind of market atmosphere developed. Interest groups promoted their viewpoints and sometimes their literature as well; business meetings occasionally erupted into rowdy debate.

In many eyes geography never looked more lively, healthy, and productive than it did at these fermenting annual meetings of the 1970s. But some of the disadvantages were all too obvious to others: even the most diligent participant during a 22-session meeting would miss 21 sessions while attending one. No *Proceedings* could possibly contain a reasonably complete and accurate record of such a meeting's events. And while the number of registered geographers grew to unprecedented levels, so did another group of participants. . .the non-paying gate-crashers. When the Milwaukee (1975) annual meeting set an all-time attendance record of 1,879 registrants, there were another 600 to 700 persons who attended but failed to register.[31] Concern over this practice revealed another problem relating to the annual meeting during the 1970s: their rising costs.

Annual meetings are costly operations for several reasons. The production, printing, and distribution of the program alone constitutes a substantial expense, and costs are incurred also in printing, mailing, and processing of registration papers, the preparation of field trip guidebooks, meetings of program and local arrangements committees, equipment rentals, security arrangements, and other requirements. It is the Association's objective that annual meetings should be self-supporting, but the demands of the large meetings of the 1970s produced several substantial deficits. The Milwaukee meeting's deficit was approximately $8,500, a figure that would have been made up if the non-registrants had paid—but it seemed impossible to ensure total registration. Hence the Council repeatedly concerned itself with cost-cutting of annual meetings, action that had the effect of ending the *Proceedings* series in its original form with the 1976 volume. Abstracts of all papers presented at each subsequent annual meeting are now published in a volume issued for each meeting. Since annual memberships do not support annual meetings operations, it was necessary also to raise registration fees repeatedly to meet rising costs. There were those members who felt that, had the programs been kept to reasonable limits, the Association would not have faced the difficulties created by the explosive annual meetings of the 1970s.

Changes in Administrative Structure

In response to the new demands of the late 1960s, the Association's administrative structure was changed significantly, so that the challenges of the 1970s could be more effectively met. First, the AAG Council on August 9, 1969 recommended to the membership a constitutional change that would replace the title of "Executive Secretary" with "Executive Director." Since the Council also included an elected Secretary (serving a three-year term), some confusion was thus eliminated. More significantly, the administrative and management functions and responsibilities of the Executive Secretary in dealing with AAG affairs were basically those of a Director—the title used by many other professional associations. This constitutional change was approved, and J. Warren Nystrom henceforth was identified as the Association's Executive Director.

Second, Salvatore J. Natoli was appointed to the Central Office staff in August, 1969, under the title of Educational Affairs Director. By late 1969, the AAG was involved in the High School Geography Project, the Commission on College Geography, the Visiting Geographical Scientist Program, the Consortium for Teacher Education (CONPASS), the Commission on Geography and Afro-America, and the Remote Sensing Project. Natoli assumed a central role in the guidance of the education-related projects, and his experience at the United States Office of Education proved invaluable. As noted earlier, Salvatore Natoli also took charge of the *AAG Newsletter.*

Third, the Central Office staff was reorganized to adjust to the new demands. Elizabeth Beetschen, who had been appointed by Executive Officer Arvin Hahn in 1963, now became Office Manager. The efficient operation of the Central Office can be substantially attributed to Elizabeth Beetschen's many years of dedicated service; she still holds the position today. In addition, specific assignments were made in areas of fiscal management, project-related editorial work, and general operations, all of which augured well for further efficiency. These moves were necessary in large measure because the Central Office workload continued to grow as the membership increased and because the Association's national office became the headquarters for projects formerly centered at universities elsewhere. And the new fiscal and personnel policies developed by the Executive Director for all AAG projects, required careful and continuous monitoring.

8. *The Association's Central Office Staff, November 1978. Front, J. Warren Nystrom, Executive Director. Rear, left to right, P. Elizabeth Beetschen, Administrative Assistant; Teresa Mulloy, Educational Affairs Assistant; Patricia McKenna, Fiscal Manager; and Salvatore J. Natoli, Educational Affairs Director. (Photograph) by Dorothy Nicholson)*

A New Central Office

Still another change involved the location of the AAG Central Office itself. After accommodating the Association, free of charge, in its national headquarters for several years, the National Geographic Society in late 1969 asked the AAG to vacate its four-room 1500 square foot facility. Executive Director Nystrom, who had been saving Association funds in lieu of rent for some time, now launched a drive to supplement these savings (plus a generous NGS grant) with sufficient money to purchase a building that would be Association-owned. The Council approved the concept of purchase in February, 1970, and the Executive Director sent letters to the membership outlining the Association's needs and objectives.[32] Describing the situation in the August-September 1970 *Newsletter,* Nystrom indicated a need for $45,000 to augment available funds in the purchase of a suitable building, among several already under scrutiny.

Initially the building fund drive progressed but slowly. On August 26, 1970, J. Warren Nystrom during the Annual Business Meeting in San Francisco was forced to report that a mere 6 percent of the members had responded to his letters of appeal, and that contributions received amounted to $18,000. A motion from the floor, proposing a building fund assessment against each full Association member, was thereupon approved without dissent.[33] As a result (and because additional members sent voluntary contributions in excess of the $10 assessment) the Executive Director could report to the Executive Committee on January 29, 1971, that the goal of $45,000 had been reached and that a new national office building had been identified.

Thus the Association acquired the address that has become familiar to members during the 1970s: 1710 Sixteenth Street, N.W., Washington, D.C. 20009. The four-story brick and stone town house, with an English basement, required considerable renovation to create eight offices, a mailing and duplicating-room, a conference room, kitchenette, ample storage space, and a small apartment to be occupied by maintenance staff. On April 24, 1971 the new facility was occupied (the third floor offices were left vacant for rental to a prospective tenant) and the June-July 1971 *AAG Newsletter* carried a series of front-page photographs by Robert D. Hodgson marking this event. The purchase of the Central Office building, viewed with concern in 1970, has proved to be an excellent investment.

9. *The AAG Central Office Building, 1710 Sixteenth Street, N.W., Washington, D.C.*

Task Forces and Project Development

Although the fiscal condition of the Association remained sound even during the period of acquisition and renovation of the new headquarters building, it was evident that long-range planning for continued funded research was an urgent necessity. The AAG's involvement in funded projects such as HSGP and CCG produced numerous benefits to the profession as well as the Association. Central Office staff was maintained in part by overhead monies derived from grant allocations and project-related publications such as the Resource Papers of CCG emanated from the Central Office. But it was evident that a major effort would have to be mounted to sustain the momentum of the late 1960s. In August, 1970 the AAG Council appointed an *ad hoc* Committee on Development and Planning that made immediate recommendations for the creation of seven Task Forces to address salient areas of geographic research and education.[34] Late in 1971, following an effort of unprecedented intensity, six of the appointed Task Forces were able to submit proposals to the National Science Foundation (the seventh was withheld to await action by the Federal Government on support for programs on marine resources).[35] And in April, 1972, the National Office was able to report that two of the Task Force proposals had been funded: the *Spatial Analysis of Progress Toward National Urban Goals in Metropolitan Areas* and *The Actual and Potential Role of Existing Black Towns as Providers of Alternative Residential Environments and Enhanced Economic Opportunity.*[36] These two projects came to be known as the "Metropolitan Analysis" and the "Black Towns" projects, and together they constituted a major element of the AAG research effort during the 1970s.

The Black Towns Project was directed by Harold M. Rose, and funding for this venture exceeded $80,000 during the period from 1972 to 1975. The project dealt with the role of existing suburban black towns as alternative residential environments for the metropolitan black population. A number of these communities were evaluated as viable residential alternatives to the inner city. The project attempted to explore the possibilities presented by non-central locations and smaller residential areas in ameliorating both social and economic problems confronted by black urban residents. It also concerned itself with the redevelopment potential of the black town—the possible transformation of these peripheral metropolitan communities into socially, economically, and aesthetically attractive places.

The Metropolitan Analysis Project, aided by nearly a half million dollars in funding, had a primary objective in assessing the progress being made toward meeting human needs in America's 20 largest metropolitan areas. It accomplished much more. In 1976, Ballinger Publishing Company published two of the three volumes emanating from the project. The first, *Contemporary Metropolitan America: Twenty Geographical Vignettes,* consists of a collection of 20 monographs (some also published separately) focussing on the cultural, political, social, and economic geography of twenty U.S. cities. The second, *Urban Policymaking and Metropolitan Dynamics: a Comparative Geographical Analysis,* provides a practical, decision-making perspective to the data gathered during the tenure of the project. And the third volume, published by the University of Minnesota Press constitutes *A Comparative Atlas of America's Great Cities: Twenty Metropolitan Regions.* The progress and products of the Metropolitan Analysis Project, directed by John S. Adams and Ronald F. Abler, immediately received critical acclaim, including review in *Mosaic,* the organ of the National Science Foundation.[37]

In the Metropolitan Analysis and Black Towns projects the Association performed exactly the kind of facilitating and coordinating role described in the *Newsletter* of January 1971. Officers and members of Council identified professional opportunities and needs, the AAG coordinated the project development effort, and the Central Office aided in administrative and fiscal areas once the projects were funded. As in the case of HSGP and CCG, the profession benefited in a multitude of ways, not least through the publication of significant new geographic literature but also by the involvement of many geographers in the work of the commissions. Similar benefits derived from the project to develop a *Sourcebook of the Environment,* funded at $83,000 by the U.S. Office of Education (1974 to 1976) and published by the University of Chicago Press in 1978; the Teaching and Learning in Graduate Geography (TLGG) Project, supported at over $250,000 from 1973 to 1978 by the National Science Foundation; and a U.S. Office of Education funded Environmental Training Survey (1974 to 1975), coordinated by Arizona State University. (An earlier project, supported by the United States Geological Survey, extended from 1973 to 1975 and involved the development of a land-use classification manual). But in the middle and late 1970s, the availability of project funds declined and the pressure on more limited funds increased. Officers of the AAG as well as Central Office administrators recognized the need for intensified project development efforts if the level of activity to which Association and profession had become accustomed was to continue.

Long Range Planning

Notwithstanding the evident need for a task force effort of dimensions similar to that of 1971, it proved difficult for the Association to mobilize the profession once more on the same scale. The Council could not agree on priorities; the appropriateness of some potential project proposals was occasionally questioned. As early as 1973, the Chairman of the AAG Project Development Committee appealed to the membership to submit proposals. In subsequent years similar attempts to generate such activity were made by the President as well.[38] But although several short-term renewals of existing projects were funded, and support was secured for a project on Soviet Resources, there was nothing in 1978 to rival the significant successes of a decade earlier. In a time of retrenchment and reduction, geographers nearly everywhere saw their research opportunities decline, their options diminish, and their educational workloads increase. Inevitably the Association felt the impact.

The effect of these changes could also be seen in the AAG membership rolls. Having passed the 3,000 mark in 1964, Association membership approximately doubled during the next five years. The *1970 Directory* contained a roster of 6,748 names. Then the total remained approximately at the same level for five years, reaching 7,072 in 1973, the peak year. The following year saw a decline of just 1 percent (student membership dropped quite significantly) to 6,994. From 1973 to 1977, the membership declined by about 6 percent.

These realities—reduced funding for fewer AAG projects and decreasing membership—were among the reasons for the Council's decision, in April, 1977, to appoint a series of Task Forces to confront Association concerns.[39] These Task Forces addressed the following areas: membership and its characteristics, services to the profession, finances, communications, governance, and management. The chairmen of the Task Forces were appointed to the AAG's Long Range Planning Committee. This Committee's first action was to disseminate, through the medium of the *Newsletter,* a questionnaire addressed to the membership in order to survey the composition of the current Association as well as members' opinions and attitudes. Of nearly 7,000 questionnaires distributed, well over 1,000 were returned; the average respondent was born in 1937 and joined the AAG in 1966.[40] Satisfaction with the AAG was expressed as somewhat stronger than moderate, and overall, the membership favored an aggressive Association. When asked to rank the publications and services of the AAG in terms of perceived usefulness, the respondents' leading choices (from 14) were (1) the *AAG Newsletter*, (2) *Jobs in Geography* and *Placement*, (3) *The Professional Geographer*, (4) Annual Meetings, (5) the *Guide to Graduate Geography Departments*, (6) the *Membership Directory*, and (7) the *Annals*. The members desired a stronger job promotion effort in business and government, stronger liaison with industry and commerce as well as government, and a greater effort in the area of academic positions. Further, members expressed a preference for limiting appearances at annual meetings to one per participant and a return to the screening of submitted papers, thus to achieve a more restricted and superior program. Such sentiments could also be seen from the respondents' desire to reduce the size of the Association's Council, although a substantial number of members expressed the belief that the Council's present size is appropriate.

Supported by the data from the survey questionnaire, the Task Forces tackled the Association's problems by recommending a number of changes to the Council, including modifications in Council membership, the formation of Specialty Groups, stream-lining of committee structure, Central Office staffing, and fiscal policies including membership fees, as well as new initiatives in the AAG's relations with other organizations, recruitment of new members, project development, and public relations. Throughout the tenure and work of the Long Range Planning Committee and its Task Forces it was evident that the Association in the late 1970s was at a crossroads: a period of continuity in the Central Office was coming to an end with the announced 1979 retirement of its long-term Executive Director in addition to all the other changed circumstances. The LRPC also considered the possibility that the AAG might dispose of its headquarters building and join a group of seven other professional associations in a common facility elsewhere in Washington.

The Constitution in the 1970s

During the turbulent 1970s the AAG Constitution and Bylaws proved adequate and durable. It was no surprise that the stresses arising from the Vietnam War and ideological differences among segments of the membership would lead to efforts to change the Association's ground rules. Among the first such attempts was a petition by Erich Isaac, representing a number of other geographers. This petition was intended to change the Constitution in such a manner that resolutions and motions adopted by Council or proposed at Annual Business Meetings "must fall within the scope of the objectives of the AAG as stated in Article II of the Constitution; those outside the scope of these objectives are to be ruled out of order."[41] A related proposal was to modify Article VII so that one hundred members, not just 25, would have to sign a petition to initiate constitutional reform. Further, a new Section 9 was to be added to the Bylaws, stating that, should 50 members of the Association consider that a resolution adopted by Council or the Business Meeting is unrelated to AAG objectives, the officers of the Association must submit that resolution's appropriateness as a mail referendum to the membership. In a mail vote held in February, 1973, the membership approved these constitutional changes, along with other, more routine modifications.[42] The results of this vote had some significance, because the Council had expressed itself as opposed to the "scope of objectives" resolution on grounds of unnecessary duplication. However, only 1,616 of over 6,000 distributed ballots were returned. As J. Warren Nystrom reported to the Council on April 13, 1973, the issues appeared to have confused many members.[43]

In 1975, the Council again addressed the Constitution, making several administrative and operational changes. For example, it was stipulated that resolutions to be presented at the Annual Business Meeting must be posted in a conspicuous location by the Secretary, at least 24 hours before the Business Meeting commences. Another change involved the Association's Regional Divisions (Article IV, Section 1): henceforth each Regional Division shall elect *by mail ballot* one Councillor from that Region. This had not always happened, and the change ensured consistency. Also, it was stated that the terms of Association Officers would begin on July 1 following the Annual Meeting rather than (as the Constitution had previously stipulated) on the day following the Annual Meeting. The Constitution might occasionally be found wanting in minor administrative areas, but its broad outlines remained intact.

The Regional Divisions

The Association's nine Regional Divisions mirrored the growth and well-being of the AAG during the 1963—1979 period. On July 31, Division membership of July 31, 1977, is revealed in the diagram on page 174.[44] The total Regional Division membership of 5,915 does not exactly match Association membership because a geographer may join an AAG Regional Division but not the national organization itself, and vice versa.

Regional Division	Total Members
Pacific Coast	1,090
West Lakes	975
East Lakes	930
Southeastern	626
Middle Atlantic	576
Middle States	564
Great Plains/Rocky Mountains	467
New England/St. Lawrence	347
Southwestern	340

In 1963, the Pacific Coast Division celebrated its 25th anniversary, and several other AAG Regional Divisions also have a lengthy history. Annual meetings, became large and well-attended; in 1967 the Pacific Coast Division and West Lakes Division annual meetings for the first time required concurrent sessions to permit presentation of all accepted papers. Concurrent sessions are commonplace today at the annual meetings of other Divisions as well; programs include field trips, plenary sessions, frequently a banquet, and other elements of a full-scale annual convention of a professional association. Occasionally, adjacent Regional Divisions organize a joint annual meeting, as was done by the Southeastern and Middle Atlantic Divisions in November, 1976.

Perhaps the strongest evidence of the vitality of the Association's Regional Divisions during the 1963—1979 period was in their publications. The *Southeastern Geographer,* first published in 1961, has become one of the profession's most respected regional journals. The *Great Plains-Rocky Mountain Geographical Journal* was founded in 1971 to carry the Division's Annual Meeting papers as well as other scholarly publications. The *East Lakes Geographer* was launched in 1965. Oldest among the regional journals is the *Yearbook of the Association of Pacific Coast Geographers,* published annually since 1935. Members of the Southwestern Division have been represented by a geographer on the editorial board of the Social Science Quarterly. One of the active Divisions that had not produced a journal, the Middle Atlantic Division, did so in 1975, when the first issue of *The Middle Atlantic* appeared. The Middle States Division began publishing a volume containing papers presented at its Annual Meetings, the *Proceedings of the Middle States Division,* in 1966. Annual published *Proceedings* have been instituted by the New England—St. Lawrence Valley Division. Papers read at Annual Meetings of the West Lakes Division have at times been published in the *Bulletin of the Illinois Geographical Society.*

The Regional Divisions in some cases have assumed a particular character, pursued specific interests, and represented certain constituencies in the Association as a whole. The Middle Atlantic Division, for example, has strong representation from geographers in government and other non-academic fields. The views of this sector of the AAG's membership have been forcefully stated by the regional Councillors from the Division. The Southeastern Division's Southern Studies Committee has promoted geographic research in and on the Southern United States. The *Yearbook of the Association of Pacific Coast Geographers* carries the impress of the school of cultural geography that centers on geography departments of the West. From all points of view, the cohesion and productivity of the Regional Divisions strengthened the Association as well as the profession during the period 1963 to 1979.

External Relations of the AAG

The Association's relations with other organizations during the 1960s and 1970s ranged from close and productive cooperation to ephemeral and ineffective communication. During the Cohen administration (1964-65) the Association's contacts and cooperation with the United States Office of Education was strengthened greatly. Executive Director J. Warren Nystrom made it one of his responsibilities to establish ties with other government agencies as well, notably the National Science Foundation where Howard H. Hines was supportive of AAG initiatives. On the Washington scene the AAG never had greater visibility. Nystrom was elected Chairman of the Conference of Secretaries of the American Council of Learned Societies; he also played a leading role among the eight Directors of the Conference of Social Science Associations (COSSA). The Central Office assiduously maintained its contacts with geographers in responsible positions throughout the Washington area, on the principle that an effective presence might generate advantages for the profession at large.

In many respects, the Association's relations with other geographical organizations were less successful. Although a member of the AAG Council became Director of the American Geographical Society and another Council member was Editor of the principal journal of the National Council for Geographic Education, cooperative efforts among the three national organizations failed more often than they succeeded.

Overtures repeatedly were made to the National Council for

Geographic Education with the ultimate objective of a merger. Some cooperation with NCGE had taken place, for example in the context of CONPASS' Committee on Education that included Ann E. Larimore and Merrill K. Ridd from AAG and Benjamin Richason III and John M. Ball from NCGE.[45] Simultaneously a "Committee of Six" was appointed "to study ways and means of closer association and cooperation between the two geography professional associations." AAG representatives were John R. Borchert, Edwin H. Hammond, and J. Warren Nystrom; from NCGE came Lorrin G. Kennamer, William D. Pattison, and Karl A. Robert. AAG Council in 1970 appointed Alvin A. Munn as a seventh member of the Committee and NCGE appointed Robert E. Gabler as the eighth member.

Nothing productive came from the discussions of the AAG-NCGE Committee. On August 21, 1971, the Council "authorized the Joint Committee to formulate a constitution and by-laws of a new unified organization," to be referred to the governing boards of AAG and NCGE; if approved, ratification would be by two-thirds of the vote of both memberships.[46] This proved to be an optimistic assignment. The Joint Committee failed to make progress and eventually it became inactive. Then, on April 21, 1972, the Council was alerted to the reality of NCGE's financial difficulties. In response, the Council approved a motion that "to assist a geographical organization during a period of financial stress, the AAG offer the NCGE free rent for two years in the AAG Central Office Building for the NCGE Central Office with possibility of extension, and to offer $1,000 to aid with expenses of the NCGE office move.[47] On June 3, 1972, J. Warren Nystrom reported to the Executive Committee that NCGE was in receipt of the offer but had deferred action. Shortly thereafter, the President of NCGE declined the AAG's offer but, at the Chicago AAG Council Meeting of November 3, 1972, the NCGE's Executive Secretary appeared to request an AAG loan to NCGE. The Council agreed only to purchase the NCGE's bond holdings, at par value, not to exceed $8,000. Even this conservative action did not gain unanimous Council support.[48] In the meantime, a merger proposal was placed before the NCGE membership, but failed to attract sufficient support. When, in 1977, the NCGE's Central Office moved from Chicago to the University of Houston, it appeared that a decade of desultory negotiation had ended without results. In 1978, however, the AAG Executive Committee scheduled its fall meeting in Milwaukee to coincide with the NCGE's Annual Convention, and as a result dialogue leading toward closer cooperation of the organizations was resumed.

In the early 1960s the other national geographic organization, the American Geographical Society based in New York, was in sound condition. *The Geographical Review,* the major AGS publication, was viewed by many geographers as the best geographic journal in the country; *Current Geographical Publications* remained a widely-used research source; and monthly issues of *Focus* provided valuable, up-to-date materials on cities, regions, and geographic ideas. For many years contacts between the AAG and AGS were close and effective; the Executive Director of the AAG served on the Council of AGS. David Lowenthal, cultural geographer on the AGS staff, served on the AAG Council. Liaison between the AAG and the AGS, the oldest professional geographical organization in the United States, remained mainly a matter of informational exchange.

All this changed during the 1970s. After years of absence from Council deliberations, the AGS was discussed in 1972. The first hint of trouble is recorded in the Minutes of the April, 1972 Council Meeting in Kansas City, when Council approved a motion "that the AAG offer the AGS a free full-page advertisement in any AAG publication to promote the sale of AGS publications."[49] This motion followed a report on the American Geographical Society by J. Warren Nystrom that detailed some of the emerging problems facing AGS—not least among them the deterioration of the neighborhood in which the AGS facility in New York was located. Several AAG Officers and Councillors in one way or another came to the aid of AGS, including Vice President Julian Wolpert and Councillor Robert B. McNee, the latter as the Society's new Director. During the fall of 1974 President James J. Parsons visited McNee to discuss ways in which the AAG might aid the AGS, and later wrote a letter published in the *New York Times* urging assistance to the Society. But AGS' limited membership, rising costs, and minor project funding combined to force a drastic cut in staff and the elimination of services. On October 25, 1976, AGS Director McNee reported to the AAG Executive Committee that the AGS Council had voted unanimously to leave New York and to occupy premises on the campus of the University of Wisconsin-Milwaukee. But this move became entangled in legal issues in the State of New York, and AGS Director Sarah K. Myers, who succeeded McNee, told the AAG Council on April 26, 1977 that the success of a new five-year survival plan, developed with AAG assistance, depended on a smooth and efficient relocation. Voices were raised in the Association's Council that urged a plan for the salvation of *The Geographical Review* and *Current Geographical Publications,* as well as the unparalleled AGS Library, should the legal impediments destroy the Society before its Milwaukee move could be consummated. In the summer of 1978 legal problems of the AGS were resolved. The library was moved to Milwaukee but the Society and *The Geographical Review* remained in New York.

The Association's liaisons with the National Academy of Sciences-National Research Council, the National Council for Social Studies, the American Association for the Advancement of Science, the American Council of Learned Societies, the Social Science Research Council and other academic and governmental offices strengthened during the 1960s and 1970s, often with salutory results. AAG representatives also serve on the United States Committee for the International Geographical Union, including the Executive Director. In international spheres, the AAG has been instrumental in the organization, funding, and implementation of several so-called Binational Seminars, in which NSF funds are applied to exchange programs between United States geographers and colleagues from foreign countries. The Soviet Seminar and the Hungarian Seminar were funded in 1975, and other Binational Seminars were being developed during the late 1970s.

Publications of the Association, 1963-1979

An observer of AAG affairs once said to the Council that the geographic profession's greatest asset was its productive and effective Association, and that the Association's greatest asset was its excellent publication program. And indeed, the membership has come to expect much from the AAG in the publications area: four issues of the prestigious *Annals,* four issues of the informative *Professional Geographer,* and ten issues of the varied *Newsletter* per year. In addition, members receive, free of charge or at a substantial discount, volumes of the *Guide to Graduate Departments,* the *Directory,* the *Monograph Series,* and published products of AAG projects, such as the very useful *Resource Papers* of the Commission on College Geography.

In the mid-1960s the Association's publication program was comparatively modest: the *Newsletter, Guide,* and various project publications had not yet begun, although *Careers in Geography* was in preparation and a *Handbook Directory* underwent revision. The Council concerned itself with problems involving the size of *Annals* and *Professional Geographer* printing runs, problems that arose from the rapid increase in membership being experienced during this period. In March, 1964, the AAG Publications Committee (Evelyn L.

Pruitt, Chairperson, Robert S. Platt, Hallock F. Raup, Thomas R. Smith, Joseph E. Spencer) presented a proposed Publications Policy stipulating, for all serial AAG publications, guiding principles, editors' responsibilities, Council responsibilities, and format.[50] Thus the *Annals* was to "consist of various articles, each starting with a brief abstract, reviews, and abstracts of papers presented at annual meetings. . .the *Annals* may contain letters, notes, committee and other reports, symposia, and other types of materials. Upon approval of the map supplement editor, with concurrence of the *Annals* editor, a special map or maps may be included." As for *The Professional Geographer,* it is characterized as "the Association's conveyor of all manner of news that is of interest to geographers. . .(it) also provides a means of conveying information in the form of very brief papers on research. . .(and) serves as a forum for the free exchange of opinions and comments, and as an outlet for *provocative satire and baiting which form part of the activities of any lively association.*" (italics added). In addition, the Publications Committee describes the objectives of the AAG *Monograph Series,* in those years a cooperative venture between Rand McNally & Co. and the Association "to solve a long-time difficulty of finding outlets for monographic studies."

By the beginning of 1965, Joseph E. Spencer had taken over as *Annals* editor from Robert S. Platt, Hallock F. Raup continued as editor of *The Professional Geographer,* and Marvin W. Mikesell had replaced Thomas R. Smith as editor of the *Monograph Series.* Individual issues of the *Annals* increased in the number of pages, and on April 14, 1967, the Council authorized Editor Spencer to employ a half-time secretary to assist in the preparation of the approaching volume of about 800 pages. Still the Editor was compelled to report to the Council on August 22, 1968, that the influx of good articles was substantial and that an 18-month backlog existed, requiring either a still larger average issue or an increase of numbers per volume. *Professional Geographer* Editor Hallock F. Raup, on the other hand, could report that the newly instituted *Newsletter* had taken space pressure off the P.G. and that the backlog was minor, amounting to two issues (four months).[51]

With the (albeit temporary) appearance of the *Proceedings,* it became evident to the Council that the 1964 Publications Policy required revision. Early in 1969, Publications Committee Chairman Lawrence M. Sommers presented a modified publications policy to the Executive Committee, and the Committee proposed that the *Annals* and *The Professional Geographer* be merged, commencing in the Spring of 1970. The Council, however, could not reach a decision during its April 1969 meeting in Detroit, but the Council did stipulate a new and more tightly formalized set of responsibilities for the AAG Publications Committee. In August, 1969 at Ann Arbor, the Council expressed itself in favor of the idea of an *Annals—P.G.* merger but insisted upon an expression of opinion from the membership at large. Following the Annual Business Meeting, the Council tabled the proposal, and in Chicago in February 1970 the Publications Committee was instructed to poll the membership by mail in connection with the future of *The Professional Geographer,* still edited by Hallock F. Raup. The results of this referendum showed that 60 percent of the membership preferred to continue the *P.G.,* and so the merger proposal was abandoned. The Council did decide, during its August, 1971 meeting in San Francisco, to reduce *The Professional Geographer's* frequency of appearance from six issues to four per year but with no reduction in the annual number of pages.

While the Association was preoccupied with the merger issue, another problem developed in the publications field. The *Monograph Series* showed few signs of life, and Rand McNally, with whose cooperation the Series had been sustained, wished to extricate itself from the arrangement. The *Monograph Series* question had several dimensions, for Editor Mikesell reported that geographers were submitting very few publishable, monograph-length manuscripts. The Council opined that Rand McNally had not been sufficiently vigorous in its advertising and promotion campaign. When an opportunity appeared to arise that might lead to joint *Monograph Series* publication between the AAG and the Institute of British Geographers, the Council instructed Mikesell to investigate. But this plan proved impractical, and the Council, concerned with rising publications costs, announced in *The Professional Geographer* of May, 1972 that no funds were available for the publication of future monographs.

The Association decided to purchase the remaining monographs in stock from Rand McNally and to market them through promotion in AAG journals. Also in 1972, David Ward was appointed to succeed Marvin W. Mikesell as *Monograph Series* Editor, although there was little prospect of activity for the Series. The Council's frustration with the situation is reflected by a statement by James R. Anderson, Chairperson of the Publications Committee, quoted in the November 3, 1972, *Minutes:*[52]

> . . .It was moved by Anderson that the Monograph Series be revitalized. He moved that funding be provided during 1973 to publish one Monograph in a newly constituted Monograph Series. The Monograph Series, he said, is in

limbo. Explorations with IBG about a joint series have not resulted in anything firm. We have an Editor, and we cannot expect him to sit and wait
upon decisions while having to hold or decline manuscripts. . .

Council voted 17-1 to support funds for one Monograph, and
David F. Ley's *The Black Inner City as Frontier Outpost* was published in 1974. But the high-quality, methodologically innovative
volume did not sell as well as the Association had hoped, and once
again the Publications Committee, in April, 1976, recommended to
Council that the *Monograph Series* be put in abeyance, a victim of
rising costs and dwindling resources.

Increasing prices for paper, binding, and mailing also plagued the
Annals, but other problems emerged as well. John Fraser Hart succeeded Joseph E. Spencer as *Annals* Editor in 1970, and in a
memorable report to the Council on April 27, 1972, Hart described
the paucity of acceptable submittals reaching his office and the consequently "thin" March, 1972 issue, then just off the press. Hart opined
that a 200-page average per issue was unrealistic, given the flow and
quality of available material; 150 pages would be more appropriate.
Indeed, the average page length in the 1970s was much less than it had
been during the 1960s, when Spencer had described 200 pages per issue
as insufficient. Again in 1974, Hart addressed the Council, urging
support for his campaign for better papers: "Young geographers do
not seem to be writing enough," he said. Among the Editor's problems also were multiple submittals—authors sending manuscripts
simultaneously to several journals. It was evident that the *Annals* had
been caught up in the ideological quarrels that afflicted the Association in the early 1970s. Councillor David W. Harvey countered Hart's
position by arguing that the *Annals'* review time was too lengthy and
acceptance criteria too conservative. He had made multiple submittals
himself, he admitted, for those reasons.

Nevertheless, the Association's *Annals* in the 1970s reflected
more accurately than ever before the variety and vitality of the
discipline. Gerard Rushton and John Hudson developed the book
review section into a valuable chronicle of the profession's salient
literature, and there was sometimes spirited debate in the commentary
section. John Hudson succeeded J. Fraser Hart as *Annals* Editor in
1976.

Although *The Professional Geographer* was relieved of its
current-events dimension by the AAG Newsletter, and in 1976 the
Council decided to transfer the Association's *Minutes* to the *Newsletter.* as well, *The Professional Geographer's* lively, informative format

was little affected. Under the editorial hand of Donald J. Patton, the *PG* from 1972 to 1978, became perhaps the most popular and widely-read professional journal the profession has ever had. High-quality concise articles, a full and representative review section, annual listings of completed theses and dissertations, productive exchanges, occasional viewpoint statements, and numerous informational items render *The Professional Geographer* virtually indispensable to the practicing geographer. On April 14, 1976, the Publications Committee recommended to the Council "that The Professional Geographer be clearly viewed as a journal with a role to support the efforts to increase the participation of non-academic (professional) geographers in the work of the Association." It could be no surprise that such an assignment would go to *The Professional Geographer,* the journal that had proved capable of virtually any role in the profession's interest.

A newly found interest in the history of the Association and concomitant interest in an archival holding revealed itself. Geoffrey J. Martin wrote "Geographers and Archives: A Suggestion"[54] which, several years later, encouraged the creation of both an official archival repository, and the Committee on Archives and Association History. Herman R. Friis, William A. Koelsch, and Geoffrey J. Martin served successively as chairman of this committee. Ralph E. Ehrenberg was appointed Association Archivist. James R. Glenn, Senior Archivist in the Smithsonian Institution, functions as curator of these papers. Committee and archival deposit have been significant developments for the compilation of this "history" of the Association. Support from the Committee has been helpful to Maynard Weston Dow for his unique project "Geographers on Film." Maynard Dow has filmed over eighty geographers in an audio-visual medium and so preserves some of the makers of the field in a manner nowhere else available. The first of his films shows Preston E. James interviewing Carl O. Sauer. The Committee has also decided to publish a "History of Geography Newsletter."[55]

Thus the AAG's publications mirror the triumphs and tribulations of other Association endeavors during the period from 1963 to 1979: unprecedented development and expansion, the inevitable problems of rapid growth, rising quality but spiraling costs, and limited resources in a stabilizing economy. But in its publications more perhaps than in any other sphere, the Association evinced its real value to the geographic profession.

The Association: 1978-1979[56]

At an earlier time in its history, the AAG responded to communications problems caused by the time and cost of travel by creating its Regional Divisions. Today, the communications problems are those resulting from scale and diversity. With a membership of over 6,000, and annual meetings open to all, the difficulties have become those of sustained and focussed dialogue. And demands for broader participation have once again called into question the limited exclusivity of the AAG's specialty committees.

Diagnosing these difficulties during its 1977-78 long-range planning process, the AAG Council has responded by enacting enabling provisions permitting the recognition of Specialty Groups within the Association, defined as "Groups of its members who share a common substantive interest in a systematic/topical specialty or in a major region." The objectives of such Specialty Groups include, but are not limited to, the organization of sessions and/or workshops for national and/or regional meetings; organization of other seminars, symposia, and conferences; development of proposals for funding; publication of a regular newsletter and other publications. The program of the 75th anniversary meeting in Philadelphia reserves time for such groups to begin to organize. It is expected that subsequent annual meetings will have increasing numbers of sessions organized by those Specialty Groups.

In the Council debate it was asked whether prospective Specialty Groups should be specified from the center, or whether groups should emerge from the membership. The latter path was chosen. A petition signed by fifty AAG members will be sufficient to have a prospective group listed on the annual membership renewal form. If one hundred members indicate a wish to become group members when they return that renewal form, the group is declared active (each AAG member can join a maximum of three groups). At the next AAG meeting it must elect its officers, and at subsequent meetings must organize one or more sessions and/or workshops. Continuation is not guaranteed. Each year, every group must submit a report and work program to the Council, and success in achieving the stated objectives will be a major evaluation criterion in determining whether a group should be continued or terminated. Thus, it is hoped that continued, focussed communication and debate can be ensured within an environment of increasing scale and diversity, and more widespread participation in Association affairs.

Changes have followed in the AAG's traditional structure. It has been recommended that the elective offices of Secretary and Treasurer be eliminated to reduce Council size and that in the future these officers be elected by Council members from among their own numbers. Specialty group committees, such as Environmental Studies, Remote Sensing, Geography and Business, Geography and Government, Urban and Regional Planning, College Geography, will go out of existence in 1979 in the expectation that if sufficient interest exists they will form again as Specialty Groups. And the Standing Administrative Committees of the AAG (Archives and Association History, Committee on Committees, Constitution and Bylaws, Employment Opportunities, Finance, Membership, Project Development, Publication, and Research Grants) will be streamlined and tied more closely to Council business by the requirement that each have two Councillors as members. The Committee on Committees will be composed of the Secretary and Treasurer. The Regional Councillors will constitute the Membership Committee. A new Committee on Specialty Groups will be formed. The Committee on the Status of Women in Geography will become a Standing Committee. By these steps, a trimmer administrative structure is envisaged as the Association begins its fourth quarter century with Patricia J. McWethy expected to become its new Executive Director, and with the prospect of the new vigor that constructive organization can engender.

9

The Association of American Geographers: Retrospect

The mission of every scholarly society is, ultimately, the promotion of the advancement of knowledge. Since there can be no single definition of what is considered to be knowledge, and certainly no one way of reaching and testing it, a scholarly society should "cherish philosophical and methodological diversity within the body of ideas with which it is most concerned."[1] It seeks to initiate and maintain a competitive discussion of scholarly questions, and it welcomes into the discussion individuals with diverse opinions. The advancement of knowledge is best achieved through the unrestricted interplay of tradition and innovation. The society should provide for special interest groups among its members—facing the very real prospect that if dissident members become sufficiently frustrated they may form a new and separate professional society. That may be desirable when the dissident group departs so far from the original body that no productive discussion is possible. But in many cases such separation is the result of a failure to appreciate the importance of including diverse points of view in the discussion process.

A scholarly society can provide numerous services for the benefit of its members. Of greatest importance is the establishment of one or more periodicals, and possibly a monograph series or other special publications. Since a profession can only exist if there are opportunities for employment, the professional society should seek to enlarge the opportunity for, and variety of, such employment. In cases where the largest number of jobs were once on university faculties, it may become essential to seek additional opportunities in government agencies, or in business firms. The professional society may also maintain a list of jobs available, and a list of members who are seeking employment.

To maintain the flow of young people into the colleges and graduate schools it is important that instruction in the elementary and secondary schools should be expanded and improved. This is especially important in the field of geography because so many students enter college in the United States without awareness of a professional field of this kind. In America, geography formerly was often criticized as a memory exercise, taught by teachers with no advanced training in professional geography. In many European countries more than a century ago the same situation existed; but due to the efforts of educators, teacher training institutions began to offer adequate training in geographic concepts and methods, thus rescuing the field from a processional of cities, rivers, capes, bays, national emblems and the like. In the United States there has been a recurring cycle of concern about the teaching of geography in the schools. Conferences, resolutions, frequent proclamations of the "new geography" come to little because not enough teachers have been trained to understand geography and the nature of the geographic undertaking. To date, little progress toward the improvement of geography teaching in the schools has been recorded.

Among the numerous other services that a learned society may provide are the preparation of lists of graduate departments of geography, and a directory of professional geographers and their specialties. The society can at least monitor programs of graduate study, listing requirements, student aids, and staffs. The society can help to establish an archive of professional records, and can urge the major universities to prepare histories of their departments. In retrospect, how well has the Association discharged the mission of a learned society?

The Association: Retrospect

The Association of American Geographers has functioned as a unique scholarly society for geographers in the United States since 1904. Other geographical societies have advanced similar goals, but none has been simultaneously national in membership and venue of annual meeting, and so frequent of change in election to office. And one might argue (until 1948) none was so selective in membership.

The objective of the Association as originally stated was:[2]

> ...the cultivation of the scientific study of geography in all its branches, especially by promoting acquaintance, intercourse and discussion among its members, by encouraging and aiding geographical exploration and research, by assisting the publication of geographical essays, by developing better

conditions for the study of geography in schools, colleges, and universities, and by co-operating with other societies in the development of an intelligent interest in geography among the people of North America.

Although the wording of the Association's Constitution has been revised occasionally, and it was extensively rewritten in 1948, the intended purpose of the Association has remained. While the functions of the Association have changed little for three-quarters of a century, the problems associated with the performance of these functions have changed frequently. It is as a crucible of intellectual exchange and a forum for competitive discussion that the Association has preserved methodological and philosophical diversity without threatening the coherence of geography as a field of scholarship. With this vital function secured, the way was made safe for the Association to extend services, to help improve the teaching of geography from the earliest grades to university studies, and to work for ever-increasing respect for the field of geography, among both geographers and non-geographers, scholar and layman.

These functions the Association has discharged throughout three distinct periods in its first seventy-five years of existence. The first of these periods extended from the founding of the Association until 1923. Emphasis was placed on the scientific investigation of geographical problems and on the establishment of geography as a field worthy of scholarly practice. The second of these periods, 1923-1948, was characterized by the emergence of geography as a distinct academic discipline and by the emergence of geography departments and graduate programs. The third period, 1948-1978, witnessed the rapid development of the Association as a professional service organization, the acquisition of a Central Office, and the development of funded projects to advance the profession and to increase its service to education, research and society.

The Years of the Titans, 1904-1923

In 1904 the initial task confronting the Association was to win acceptance for geography in the school system and to gain the acceptance of geography as a field of specialized research by the scholarly community. Membership in the Association was 48 in 1904, and 130 in 1923. This figure is conservative indeed when compared with 1902 membership figures for other societies: American Geographical Society—1,310; The Appalachian Mountain Club—1,270; The National Geographic Society—2,600; The Geographical Society of California—more than 400; The Sierra Club—500; The Philadelphia

Geographic Society–480; The Geographic Society of Chicago–150; The Alaska Geographical Society–1,200; The Geographical Society of Baltimore–1,725.

The small membership of the Association was accomplished by design. Only individuals "who have done original work in some branch of geography" were eligible for membership. Davis recognized that only in this way was scientific advance possible. While he wished to encourage and assist grade-school teachers, he did not wish to have them in the Asociation. The discipline was to be fashioned by Association members, then made available to educators. Similarly Davis was not enthusiastic about membership "of the Cook and Peary variety." It was a policy born of enthusiasm and efficiency, yet one which was bound—sooner or later—to vex the disenfranchised. Such an organization belonged in the early years of the century, but as *demos* swept the land, increased criticism was levelled at the concept of an elite corps of geographers. Yet it was in the early years of the Association that some of the long-lived features of American geography were developed.

Davis elaborated a physiography, and urged his students to "develop geography proper, physiography and ontography properly combined, and not simply physiography (as I have done too much)."[3] The word ontography, apparently minted by Davis in 1902, did not last as long as the theme. Perhaps the etymology "biome" would have endured longer than the Graeco-Davisian "ontography." But it was Davis himself who urged the study of organic response to a physical environment, and who thus facilitated disciplinary evolution. The Davisian inspired movement constituted a departure from Davis's own physiography, loci of commitment had been fashioned, and a new synthesis for American geography was in the making. Davis developed a remarkable record of geographers who studied under him: those students who continued their work in geology or physical geography include Henri Baulig, John M. Boutwell, Reginald A. Daly, J. Walter Goldthwait, Frederic P. Gulliver, Thomas A. Jaggar, Douglas W. Johnson, George R. Mansfield, François E. Matthes, Walter C. Mendenhall, Philip S. Smith, and Lewis G. Westgate. Men who studied under Davis and accepted to a larger or smaller extent his urging for ontography included: Robert L. Barrett, Isaiah Bowman, Albert Perry Brigham, Alfred H. Brooks, Robert M. Brown, Collier Cobb, Sumner Cushing, Richard E. Dodge, Herbert E. Gregory, William M. Gregory, Ellsworth Huntington, Mark Jefferson, Curtis

F. Marbut, Lawrence Martin, Vilhjalmur Stefansson, Ralph S. Tarr, Walter S. Tower, and Robert DeC. Ward. This formidable list is not exhaustive. Interesting, however, is the fact that more than 90 percent of this group were accepted to membership in the Association during the years 1904-1923.

The physiography advocated by Davis and studied then adumbrated by his students dominated the geographic thinking of a large number of the early Association members. The learning, investigation, field work, and the "working up" of results for many came to be a way of life and not only a sporadic activity—typified by Dodge's remark of Ward, that "his academic work was as a sacred duty." From an intellectual adolescence Davis's students themselves were to emerge as Titans: Johnson to study coasts, Fenneman to define physiographic regions, Martin to advance glacial geomorphology, Bowman to write on the regional geography of the Andes, Ward and Huntington to reveal something of the physics of the air and its effect upon human physiology, and Brigham to establish historical geography. And there were many more of Davis's students who shared his intellectual persuasion. The point is that the Davisian model provided a common point of scientific beginning from which structured thought could evolve. A place for man on the fundament was sought in the intellectual construct. An objective ontography was in prospect of development but was somewhat obscured by the rise of concern with geographic determinism.

The competitive discussion which ensued at the annual meetings of the Association facilitated the emergence and evolution of an initial American geography which contained the properties of a scholarly discipline. The opportunity for personal exchanges, too, was a significant part of the Association's contribution. Geographers came to know each other personally and lasting friendships were formed. This augured well for the evolution of a geography, so in need of the give and take of discussion of this sort. William Morris Davis, always eager to help a younger geographer, seized the occasion of the annual meeting to encourage a promising performance. Charles C. Colby has recalled the circumstances in which Davis invited him to discuss his paper:[4]

> I had presented a paper on the historical geography of South-eastern Minnesota at the Yale meeting (1912). . .Professor William Morris Davis had invited me to have breakfast with him on the morning following my presentation and gave me pointed suggestions and criticisms.

Of personal relationships that were formed Marcel Aurousseau has written:[5]

> I attended the last joint meeting of the American Geographical Society and the Association of American Geographers, which was held in New York in the Spring of 1922. Many of those attending were accommodated in up-town hotels, and I found myself sharing a big room with. . . .Mark Jefferson, and. . . .Harlan Barrows. Bowman asked me if I were comfortable, and I told him that we seemed to be rather crowded. "Yes" he said, "I arranged that. I want you to get to know Jefferson, and Barrows to get to know you." It worked very well, as Jefferson was an easy room mate, and I enjoyed the association.

Numerous relationships which were both personal and professional were formed. These developments facilitated discussion of work both privately and following delivery of papers at the annual meeting. Late in his life J. Russell Smith wrote of his relationship with Mark Jefferson:[6]

> I never went to college and studied under Mark (Jefferson)—never was his associate in a college faculty. My sole relationship with him has been as a fellow member of the Association of American Geographers, and in this relation I met him a number of times from 1905 on.

Correspondence between members was encouraged by the personal relationships formed at the annual meetings. Intellectual exchanges would take place in this way, occasionally lasting for several months. Points of view were fashioned and exchanged. Intellectual inspiration and renewal was provided by papers delivered and by the frequently arranged "round table discussions, ("smokers" as Davis called them). Much of this thought was written in article form; authors sought suitable publication.

In 1911 the Association established its own *Annals.* Previous to that Association members sought publication of their work in the *Bulletin of the American Geographical Society,* the *Journal of Geography* or one of the geological periodicals and occasionally *The Geographical Journal.* With the inauguration of the *Annals,* Association members had their own periodical, in which non-members were not allowed publication privileges. An editor and readers made decisions on manuscripts submitted. Some were revised and some were returned to authors for revision, but rejection was rare since authors typically submitted only excellent and original manuscripts. Authors were given numerous reprints of their article published in the *Annals* which they would frequently send to non-member colleagues, both in the United States and abroad.

One other publication was initiated in 1905. We now know it as the *Newsletter.* It has been a valued instrument of successive secretaries. First of this "series" to appear was a single printed page, "Association of American Geographers: Information to Members," on June 7, 1905 by Secretary Albert Perry Brigham. Another separate printed sheet, "Preliminary Information for Members" preceded the third annual meeting of the Association (October 8, 1906). Other printed single sheets appeared at irregular intervals entitled, "Association of American Geographers," until January 10, 1911, when the first "Circular of Information" was released by Brigham.[7]

It was in this first twenty year period of the Association that special attention was accorded the role of geography in the educational process. Association members contributed articles to the *Journal of Geography,* established Association committees to facilitate the inclusion of geography in the schools, and supported the creation of the National Council of Geography Teachers. Not only by a wealth of published material—including some textbooks—did Association members help the pedagogue, but also by individual members of the Association taking teachers on field trips and showing them how to see what they observed. Perhaps the most celebrated of the field trips undertaken by a substantial number of Association members was the Transcontinental Excursion commemorating the 60th anniversary of the founding of the American Geographical Society. Of the American geographers who participated, a large percentage were members of the Association, including the Excursion leader and his three marshals, (William Morris Davis, Mark Jefferson, Richard E. Dodge, and Isaiah Bowman). The 13,000 mile seminar on wheels provided a happy opportunity for interaction among American and European geographers. It also generated publicity for geography and geographers, and many a newspaper gave the Excursion its attention. Less well known is the fact that following an address at La Crosse, Wisconsin, L. P. Denoyer (then of the Department of Geography and Geology, State Normal School, La Crosse, Wisconsin) wrote to Albert Perry Brigham:[8]

> The writer met you on the train in La Crosse last August and heard the talks on the R.R. platform. Among the things said, Mr. Davis spoke of the Society of American Geographers of which I believe you are secretary. I should like to join your society and would like to make application.

Only a few years later Denoyer, with Otto Geppert, founded what became one of the largest cartographic enterprises in the United States.

Further attention came to the Association through the services which 51 Association members provided during World War I and the Peace Conference that followed. Geographers held positions as civilians in war agencies working especially with commodity studies and boundary problems. Walter S. Tower was appointed head of the Commodities Section of the Shipping Board. He selected Vernor C. Finch, William H. Haas, George B. Roorbach, and Charles C. Colby. Harlan H. Barrows was selected to head the War Trade Board of the Division of Planning and Statistics. Barrows recruited J. Russell Smith, Ray H. Whitbeck and Nels A. Bengtson. Additionally geographers worked for the Inquiry, an anonymous organization established at the request of President Woodrow Wilson in 1917. Isaiah Bowman was soon put in charge of this undertaking, which included several geographers in its 150 or so members. From the fact-gathering and map-making efforts of the Inquiry emerged considerable data of value to the American Delegation to Negotiate Peace at Paris. Bowman was appointed Executive Officer of the Paris Delegation of Experts. Mark Jefferson was appointed cartographer, Charles Stratton and Armin K. Lobeck were assistant cartographers. Douglas W. Johnson was appointed geographer of the commission.[9]

President Woodrow Wilson was impressed with the geographers' performance. His lack of qualifications did not permit him to join the Association, but he did take membership in the American Geographical Society. The Association already included one president of the United States in its membership—Theodore Roosevelt had joined in 1915.

The Years of New Viewpoints, 1923-1948

Shortly after Charles C. Colby of the University of Chicago assumed office as Secretary of the Association in 1923, the locus of Association activity moved abruptly to the Midwest. Faculty and one-time students of the University of Chicago began to assume office and membership in the Association in larger numbers than hitherto. The search for alternatives to the Davisian construct seemed to emanate from Chicago. This was due in part to the thought of Salisbury who was opposed to Davisian concepts and who institutionalized this viewpoint in the Salisbury seminar.[10] The dispatch of Davis and determinism was hastened by Isaiah Bowman's *The New World,* in which were summarized the arrangements between nations and countries following the Treaty of Versailles, by Harlan H. Barrows', *"Geography as Human Ecology"* and Carl O. Sauer's *Morphology of Landscape.*

Additionally, the members of the geographical field conferences sought a new formulation concerning man's use of the land[11] and improvements in the methods of studying it through the detailed mapping of representative small areas. Other geographers led a variety of intellectual sorties in which man and his resources were the object of dominant concern. The chorological concept began to emerge and a trickle of regional studies became a flood in the thirties. Many of these regional studies were of North America, and American geographers became better acquainted with their own continent. For a while, Whittlesey's sequent occupance became an organizing concept for regional studies, though by the end of the thirties it had fallen from popular use.

Institutional growth was accomplished for geography. Wallace W. Atwood, President of Clark University, inaugurated *Economic Geography* in 1925. In that same year, the Society of Woman Geographers was founded, and ten years later the Association of Pacific Coast Geographers was born. Meanwhile, Isaiah Bowman had brought new staff, vigor, and a well defined program to the American Geographical Society. Geography courses and departments were increasing in colleges and universities. The forum of the Association was invaluable in creating lasting and meaningful geographic relationships between individuals. An enthusiasm was encouraged and facilitated, a dedication emerged, and geographers were made out of individuals who might otherwise have remained but teachers of geography.

In the twenties and thirties there emerged the first generation of indigenous American geographers. This was the beginning of a vitalizing point of view, whose character was later sought by John K. Wright when he inquired "What's American about American geography?"[12] Several of the men who entered the Association in the twenties became moving forces in the development of American geography: 1924—Darrell H. Davis, Kenneth C. McMurry, Robert S. Platt, and Derwent S. Whittlesey; 1925—John K. Wright; 1926—Preston E. James, Clarence F. Jones; 1927—Richard Hartshorne, Glenn T. Trewartha; 1928—Ralph H. Brown, John B. Leighly. The list is selective but it reveals to geographers half a century later, the type of minds that were being encouraged into professional juxtaposition and dialogue. These men, among others, helped develop American geography through their books, articles, and presentations before the Association. Presentations of papers at the Association's annual meetings are less visible perhaps than published books and articles. Yet they are the matrix of the Association, and around them thought and discussion flow. The value of an individual's contribution to the

Association cannot be measured solely in terms of the number of papers delivered during a life-time, but that measure does indicate persistence, and in the days before open admission of membership and papers, it necessarily meant scholarly excellence. Those geographers who presented twenty or more papers before the Association include: Wallace W. Atwood, Albert Perry Brigham, Charles F. Brooks, Robert M. Brown, Raymond E. Crist, William Morris Davis, Richard Hartshorne, William H. Hobbs, Ellsworth Huntington, Preston E. James, Mark Jefferson, W.L.G. Joerg, Douglas W. Johnson, Lawrence Martin, François E. Matthes, Robert S. Platt, J. Russell Smith, Glenn T. Trewartha, Stephen S. Visher, Ray H. Whitbeck, and Derwent S. Whittlesey. Owing to discrepancies between the listing of abstracts in the *Annals,* the printed program, the distinction between a paper and what otherwise might be termed a presentation, and the reality of what happened, exact figures concerning papers delivered are subject to review. Nevertheless, it does seem as though Mark Jefferson read more papers than any other person in the history of the Association.[13] It seems he is followed by Preston E. James. It also seems that during the 75 years of the Association the two members who have been most continuously active over the longest period are Preston E. James (1926–), followed by Richard Hartshorne (1927–). It was essentially the second generation of indigenous American geographers that dominated the twenties and thirties of American geography. There was a restless seeking and searching, an intellectual striving, an almost existential evolutionary growth towards an ultimate objective, yet one which had not yet been defined. Through these years, academic geographers were gaining rapidly in numbers, while the restrictive membership policy of the Association remained in effect. This precipitated the dramatic confrontation between the Association and the Washington-based geographers in 1943. The five years which ensued were troubled ones in the history of American geography, but with merger in 1948, a new Association, with a new constitution, apparatus, and point of view, had been fashioned.

The Years of Professionalization, 1948-1978

Arising in the post-war years of American geography was an Association with a newly acquired professional look about it. The variety of articles written began to reflect the spread and fragmentation of the field. The *Annals* continued to function as the major journal of the Association, though the membership related to it in a man-

10. Preston E. James

11. Richard Hartshorne

ner quite different from the early membership. By 1977 it could be asserted that 99 of every 100 members of the Association did not submit manuscripts to the *Annals,* and that three of every four manuscripts submitted were not accepted for publication. It was the influx of articles by geographers outside the United States who helped maintain the present size of the *Annals*; approximately one-third of the current submissions, and approximately one-third of the articles published in the *Annals,* derive from that source. This represented a stark change from the early years of the Association when only members could submit articles, and when rejection of an article constituted an unusual practice.

Larger than ever before, the Association offered more professional services and support than previously, and became purveyor of a new geographical point of view. Among new electronic devices that supported methodological change in numerous of the social sciences was the computer. Originally developed during World War II to facilitate rapid aim of anti-aircraft batteries, it was first put to non-military use by the U.S. Bureau of the Census in 1950. Remote sensing was highly developed in World War II, and with new kinds of planes, cameras and films began to emerge as a geographical tool, along with computer graphics. A movement, gathering strength in the fifties, adopting mathematical models and statistical techniques and known as the "quantitative revolution," largely spent itself in the sixties. But we may hope that the methodologies devised during that period will not be lost or abandoned. Geography also benefitted from the heightened interest in environmental problems. For the purpose of maintaining philosophical and methodological diversity geography is viewed as a field of science by some and a field of literature by others. There are insights to be gained from both approaches, and the competitive discussions among followers of these and other proposals should be preserved.

The Association became a forum for geographers of very diverse knowledge and beliefs. Qualifications for membership in the Association were reduced, until willingness to pay the annual dues was the test for entry. Papers which had been screened were no longer screened. Attendance at the annual meeting increased and a record attendance of 3,000 at the New Orleans meeting in April, 1978 was registered. Geographers seemed to specialize more and more as the scope of their undertaking seemed to extend. Disciplinal specialization often led to more than 20 concurrent sessions at the annual meetings. Special interest groups began to emerge, then evolve under the impetus of their own special interest, and threatened to split away from the parent organization. In fact the problems of the Association (1948-1978) have been shared by others of the social sciences. The major problems which have confronted professional geography in the United States during these years can be grouped under three headings: 1) joining diversity of approach with unity of objective; 2) broadening the services provided by the Association for its members, and extending its services to groups not yet adequately covered; 3) improving the teaching of geography at all levels.

Joining Diversity of Approach with Unity of Objective

The preservation of competitive discussion of the objectives, concepts, and methods of geographical study is perhaps the most important challenge to be faced. Seventy-five years ago the problem was how to start such a discussion among qualified scholars; today the problem is to keep the discussion active, with widespread participation among members of the profession. The idea is not to narrow the options, but to maintain free expression of diverse points of view.

These objectives are not easy to reach. There is a tendency to form separate groups—special interest groups they are called. The members of these groups also start a competitive discussion of their own segment of the larger field, or perhaps their own special method of study. The members of these groups tend to put their own interests above the more diffuse interests of the larger group. Eventually these groups may break apart from the parent society, hoping to form a new professional field, and in many cases these separations have been successful. This is not to suggest that special interest groups should not be permitted to break apart—but only to emphasize the need to permit free and open discussion in which as many members of the Association as possible participate.

The greatest difficulty occurs perhaps when scholars who use quantitative procedures find no common grounds to justify their continued association with those who use the literary methods. But the benefits and losses of such a separation need to be freely and mutually discussed.

The Long-Range Planning Committee (formed in 1977) has recognized the importance of encouraging and giving support to special interest groups, but recommends attempting to hold them more closely to the Association as a whole by asking the leaders of each group to prepare progress reports, and a summary of papers and discussions at the annual meetings.

Services to Members. During the years when the Association was managed by a part-time, unpaid secretary, it was impossible to provide the services that members of other scholarly societies were receiving. As early as 1913 Richard E. Dodge had discussed a "home" and "permanent officer for the Association," in the building of the American Geographical Society. During the early 1920s, when Richard E. Dodge was both secretary of the Association and editor of the *Annals,* he urged the Council to employ a full-time paid secretary.[14] But the Council was not able to provide the necessary financial support. The first paid employee of the Association was Evelyn Petshek, who was hired in 1950 to take care of mailing, mimeographing, and the maintenance of a central file of the membership. It was not until 1963 that the decision was made to employ a full-time executive secretary. Arvin W. Hahn was appointed to this position. He was soon succeeded by Saul B. Cohen, and then John Fraser Hart. Then came J. Warren Nystrom to assume office for 13 years. The Association was under spreading sail.

As the membership grew, and the Association finances became more secure, the Central Office in Washington, D.C. was provided with a staff. The Executive Director, the staff, the officers, and various committees worked together and increased the membership substantially, and also undertook to seek financial support from a variety of potential sources. Gifts and grants made it possible to provide the members with a larger range of services.

The Teaching of Geography. The Association has been concerned about the teaching of geography at all levels throughout its history. As early as 1910 the Council passed a resolution identifying inadequate teacher training as the chief cause of the failure of physical geography to win a place in the secondary schools. That a teacher of geography should have studied the subject at a higher grade level was not a very challenging recommendation. With some revision of the words, the same resolution might be passed today. It is still difficult to explain that advanced training in geographical concepts and methods—including the use and construction of maps—is essential for the successful teaching of the subject.

The teaching of geography, however, has not remained entirely unaffected by the thinking of professional geographers, owing largely to the availability of federal funding beginning in the early 1960's. Recognition of geography in the National Defense Education Act, as amended in 1964, led to the organization of teacher institutes across the nation, some of which were coordinated by the Association. In these institutes professional geographers directly taught teachers, which encouraged the opening of many other channels of communication with educators. In the same year, the National Science Foundation undertook sponsorship of the Association's High School Geography Project, an organization that brought together hundreds of people, including classroom teachers and professional geographers, for the development of instructional materials that authentically reflected modern geography. These materials, since revised, have probably enjoyed their largest success in teaching teachers as well as high school students.

The Commission on College Geography, also financed by the National Science Foundation, did not try to develop just one set of materials for undergraduate curricula, but recommended a variety of approaches, and provided resource papers for the use of college teachers.

Envoi

Seventy-five years have now passed, and history records that the Association has actively facilitated the creation and expression of a discipline, and the development of a professional field. Challenges have been posed—of making peace, planning land use, evolving a methodology, et al.; responses have been forthcoming. Growth of the scholarly field was thus accomplished.

Unity of the discipline has given way to ever-increasing diversity. From simplicity came dichotomy, from unity emerged diversity, and from diversity came pluralism. Fragmentation meant new pioneer zones, new laboratories for learning, new ways to arrange old problems, specialist tools to probe new problems. A continual striving for understanding and excellence has meant creative tension, and, provided that this is not disruptive, such a position is full of promise.

Geography, now ensconced in colleges and universities throughout the country, demonstrates vitality and unity as a servant of society. It is as a forum for discussion and as a hearth of learning that the Association will continue its good works.

NOTES

Chapter 1

Prolegomenon: The Emergence of a Field of Learning

[1]Charles C. Colby, "Changing Currents of Geographic Thought in America," *Annals of the Association of American Geographers,* Vol. 26, No. 1, (1936), p. 1.

[2]Arnold Henry Guyot was appointed Professor of Geology and Physical Geography at Princeton in 1854.

[3]John K. Wright, "The Field of the Geographical Society," Chapter 23 in *Geography in the Twentieth Century,* edited by Griffith Taylor, (1951), pp. 544-548.

[4]*The Encyclopaedia Britannica,* eleventh edition, 1910-1914, Vol. 11, p. 630.

[5]*Geography Through a Century of International Congresses,* (International Geographical Union. Commission on the History of Geographical Thought) (1972), 252 pp.

[6]Merle Curti, *The Growth of American Thought,* (1943), pp. 587-588.

[7]Paul F. Griffin, "The Contribution of Richard Elwood Dodge to Educational Geography," dissertation presented to the faculty of philosophy at Columbia University (1952), pp. 92-94.

[8]William Morris Davis, "The Progress of Geography in the United States," *Annals of the Association of American Geographers,* Vol. 14, No. 4, (1924), p. 165.

[9]Charles C. Colby, "Changing Currents of Geographic Thought in America," *Annals of the Association of American Geographers,* Vol. 26, No. 1, (1936), p. 9.

[10]John Wesley Powell, "The Organization and Plan of the United states Geological Survey," *American Journal of Science,* Vol. 29, (January-June, 1885), pp. 93-102.

[11]Hugh R. Mill, "Geography in European Universities," *Educational Review,* Vol. 6, (1893), pp. 417-418.

Halford J. Mackinder, "Modern Geography, German and English," *The Geographical Journal,* Vol. 5, (1895), pp. 367-379.

[12]For a careful report of this methodological discussion see Richard Hartshorne, *The Nature of Geography: A Critical Survey of Current Thought in the Light of the Past* (1939).

[13]John K. Wright, *Geography in the Making: The American Geographical Society, 1851-1951,* (1952), p. 382.

[14]Preston E. James, "The Association of American Geographers," *American Council of Learned Societies Newsletter,* Spring-Summer, 1975, pp. 9-18.

[15]Wilbur Zelinsky, "The Demigod's Dilemma," *Annals of the Association of American Geographers,* Vol. 65, (1975), pp. 123-143.

[16]Richard J. Chorley, Robert P. Beckinsale, and Antony J. Dunn, *The History of the Study of Landforms or the Development of Geomorphology. Volume Two: The Life and Work of William Morris Davis* (1973).

[17]Geoffrey J. Martin, "A Fragment on the Penck(s) - Davis Conflict," *Geography and Map Division, Special Libraries Association Bulletin,* No. 98, (1974), pp. 11-27.

[18]Carl O. Sauer, "The Morphology of Landscape," *University of California Publications in Geography,* Vol. 2, No. 2, (1925), pp. 19-53, reprinted in 1938.

[19]Stephen S. Visher, *Scientists Starred 1903-1943* in *"American Men of Science,"* (1947), pp. 373-382.

[20]Marvin W. Mikesell, "The Rise and Decline of Sequent Occupance: A Chapter in the History of American Geography," *Geographies of the Mind. Essays in Historical Geosophy,* edited by David Lowenthal and Martyn Bowden, (1976), p. 150.

Chapter 2

American Geography in Universities and Geographical Societies

[1]Dominic Vuolo, Jr., "The Evolution of Academic Geography in the United States, circa 1890-1925," M.S. thesis submitted to the Department of Geography, Southern Connecticut State College, (1978).

For further details see the course catalogues for each of these institutions.

[2]Richard J. Chorley, Robert P. Beckinsale, and Antony J. Dunn, *The History of the Study of Landforms, or the Development of Geomorphology, Volume 2. The Life and Works of William Morris Davis,* (1973).

[3]Ibid., p. 34.

[4]William Morris Davis and Reginald A. Daly, *The Development of Harvard University: Since the Inauguration of President Eliot. 1869-1929:* (edited by Samuel Eliot Morison), (1930), pp. 314-315.

[5]Davis was appointed Professor of Physical Geography in 1890, and Sturgis Hooper Professor of Geology in 1899. He taught at the University of Berlin in 1909, and at the Sorbonne in 1911-1912. He retired from Harvard in 1912 at the age of 62, but lived for an additional 22 years of scholarly productivity during which he held temporary appointments at Oregon, California, Arizona, and Stanford Universities, and the California Institute of Technology.

[6]John Wesley Powell, *Report on the Lands of the Arid Region of the United States, with a More Detailed Account of the Lands of Utah,* (45th Congress, 2nd Session, Washington, D.C., 1878); reprinted and edited by Wallace Stegner, Harvard University Press, (1962.)

[7]William Morris Davis, "Geographic Methods in Geologic Investigations," *The National Geographic Magazine,* No. 1, (1889), pp. 11-26.

[8]Arnold Guyot, *Physical Geography,* (1875).

[9]William Morris Davis, "Methods and Models in Geographic Teaching," *The American Naturalist,* Vol. 23, (1889), pp. 566-583; reprinted in William Morris Davis, *Geographical Essays,* 1909, pp. 193-209.

[10]Geoffrey J. Martin, "The Ontographic Departure from Davisian Physiography," paper presented at the Association of American Geographers annual meetings, New York City, April, 1976.

[11]Israel C. Russell, "Reports of a Conference on Geography," *Journal of the American Geographical Society,* Vol. 27, No. 1, (1895), pp. 30-41.

[12]John K. Wright, "Daniel Coit Gilman," *Human Nature in Geography,* Chapter 11, (1966), pp. 168-187.

[13]William Warntz, *Geouraphy Now and Then: Some Notes on the History of Academic Geography in the United States,* (1964), pp. 51-63.

[14]Geoffrey J. Martin, *Ellsworth Huntington: His Life and Thought,* (1973), pp. 70-74.

[15]Derwent S. Whittlesey, "Dissertations in Geography Accepted by Universities in the United States for the Degree of Ph.D. as of May, 1935," *Annals of the Association of American Geographers,* Vol. 25, (1935), p. 213.

[16]Emory R. Johnson, *Life of a University Professor: An Autobiography,* (1943), p. 25.

[17]Virginia M. Rowley, *J. Russell Smith,* (1964), pp. 73-74.
    J. Russell Smith, "American Geography 1900-1904," *The Professional Geographer,* Vol. 4, No. 4, (1952), pp. 4-7.

[18]William Morris Davis, "The Progress of Geography in the United States," *Annals of the Association of American Geographers,* Vol. 14, No. 4, (1924), p. 201.

[19]William Warntz, *Geography Now and Then: Some Notes on the History of Academic Geography in the United States,* (1964), pp. 150-151.

[20]*Annals of the Association of American Geographers,* Vol. 25, No. 1, (1935), p. 213.

[21]*Annals of the American Academy of Political and Social Science,* Vol. 18, (1901), pp. 446-468.

[22]"The Economic Geography of the Argentine Republic," *Bulletin of the American Geographical Society,* Vol. 35, (1903), pp. 130-143.

[23]"The Economic Geography of Chile," *Bulletin of the American Geographical Society,* Vol. 36, No. 1, (1904), pp. 1-21.

[24]*University of Pennsylvania Publications, Series in Political Economy and Public Law,* No. 17, (1905), 155 pp.

[25]"Geography and International Boundaries," *Bulletin of the American Geographical Society,* Vol. 35, No. 2, (1903), pp. 147-159.

[26]Charles C. Colby, "Ellen Churchill Semple," *Annals of the Association of American Geographers,* Vol. 23, (1933), pp. 229-240.

[27]*The Geographical Journal,* Vol. 17, (1901), pp. 588-623; reprinted in *The Bulletin of the American Geographical Society,* Vol. 42, (1910), pp. 561-594.

[28]*The Courier-Journal,* Louisville, June 1, 1903.

[29]*The Annals of the American Academy of Political and Social Science,* Vol. 25, No. 1, (1905), p. 153.

[30]"A Half Century of Geography - What Next?," A publication by the Department of Geography, University of Chicago, (1955), 40 pp.

[31]"Geography in the University of Chicago," *Bulletin of the American Geographical Society,* Vol. 35, No. 2, (1903), pp. 207-208.

[32]William D. Pattison, "Goode's Proposal of 1902: An Interpretation," *The Professional Geographer,* Vol. 30, (1978), pp. 3-8.

[33]Edward C. Hayes, "Sociology and Psychology; Sociology and Geography," *The American Journal of Sociology,* Vol. 14, (1908), pp. 371-407.

[34]A. Hunter Dupree, *Asa Gray, 1810-1888,* (1959), p. 261.

[35]John K. Wright, *Geography in the Making: The American Geographical Society, 1851-1951,* (1952).

[36]Gilbert H. Grosvenor, *The National Geographic Society and Its Magazine,* (1957).

[37]*The Physiography of the United States: Ten Monographs,* by John Wesley Powell, Nathaniel S. Shaler, Israel C. Russell, Bailey Willis, C. Willard Hayes, J.S. Diller, William Morris Davis, and Grove Karl Gilbert, (American Book Company, 1897). First issued in 10 parts in 1895 under the title, *National Geographic Monographs* under the auspices of the National Geographic Society.

[38]Gilbert Grosvenor, *The National Geographic Society and Its Magazine,* (1957).

[39]Alexander Graham Bell, "The National Geographic Society," *National Geographic Magazine,* 23, (1912), p. 274.

[40]Alexander Graham Bell to Ferdinand von Richthofen, September 15, 1899.

[41]*The National Geographic Magazine,* Vol. 12, (1901), pp. 351-357.

Chapter 3

The Formation of the Association of American Geographers, 1903-1904

[1]William Morris Davis, "Geography in the United States," *Science,* N.S. Volume 17, No. 473, (1904), pp. 121-132.

[2]Herman LeRoy Fairchild, *The Geological Society of America 1888-1930: A Chapter in Earth Science History,* (1932), xvii and 232 pp.

[3]"The Association of American Geographers, 1903-1923," *Annals of the Association of American Geographers,* Vol. 14, No. 3, (1924), pp. 110-111.

[4]Ibid., p. 111.

[5]Circular letter, William Morris Davis to Albert Perry Brigham, January 26, 1904.

[6]Ibid.

[7]Ibid.

[8]Minutes of the Association of American Geographers, p. 1.

[9]Richard E. Dodge to Isaiah Bowman, February 23, 1917.

"In the scrap book I find a copy of Davis's call for the first meeting of the Association of American Geographers. On the back is the interesting list below, showing Davis's then idea of Geography. . ."

[10]John K. Wright, *Geography in the Making: The American Geographical Society, 1851-1951,* (1952), p. 144.

[11]*Report of the Eighth International Geographical Congress, Held in the United States, 1904,* (1905).

[12]Preston E. James and Ralph E. Ehrenberg, "The Original Members of the Association of American Geographers," *The Professional Geographer,* Vol. 27, (1975), pp. 327-335.

Many of them are sketched, with reference to fuller biographies, in Stephen S. Visher, "Notable Contributors to American Geography," *The Profes sional Geographer,* Vol. 17, (1965), pp. 25-29; and Vol.18, (1966), pp. 227-229.

[13]Those who were invited but declined to join the Association as charter members include: Frederick V. Coville, William F. Ganong, Adolphus W. Greely, C. Willard Hayes, John F. Hayford, Willard D. Johnson, Frederick H. Newell, George R. Putnam, William Z. Ripley, Otto H. Tittmann.

[14]Those proposed for charter membership but opposed by one or more members of the Committee on Organization include: Cleveland Abbe, Sr., William L. Bray, Henry H. Clayton, Harold W. Fairbanks, Thomas H. Kearney, Lindley M. Keaseby, Edward M. Lehnerts, A. Lawrence Rotch, J. Russell Smith, Spencer Trotter, Alfred W.G. Wilson, Leslie H. Wood.

[15]Minutes of the Association of American Geographers, pp. 11-17.

[16]Minutes of the Association of American Geographers, p. 3.

[17]The Constitution of the Associatqon of American Geographers.

[18]Archives of the Association of American Geographers.

[19]Mark Jefferson to William Morris Davis, March 7, 1905.

[20]Minutes of the Association of American Geographers, p. 9.

[21]*Bulletin of the American Geographical Society,* Vol. 37, No. 2., (1905), p. 85.

[22]Minutes of the Association of American Geographers, p. 6.

Annals of the Association of American Geographers, Vol. 1, (1911), pp. 101-102.

[23]Minutes of the Association of American Geographers, pp. 7 and 8.

Chapter 4

The First Twenty Years, 1904-1923: A Question of Identity

[1]Charles Redway Dryer, "A Century of Geographic Education in the United States," *Annals of the Association of American Geographers,* Vol. 14, (1925), p. 148.

[2]Earl Shaw to Geoffrey Martin, November 10, 1966.

[3]Derwent S. Whittlesey, "Dissertations in Geography, Accepted by Universities in the United States for the Degree of Ph.D. as of May 1935," *Annals of the Association of American Geographers,* Vol. 25, (1935), pp. 211-237.

[4]William Morris Davis, "An Inductive Study of the Content of Geography," *Bulletin of the American Geographical Society,* Vol. 38, (1906), pp. 67-84.

[5]Nevin M. Fenneman to Albert Perry Brigham, April 20, 1912.

[6]"The Association of American Geographers, 1903-1923," *Annals of the Association of American Geographers,* Vol. 14, (1924), p. 112.

[7]Herbert E. Gregory to Albert Perry Brigham, March 9, 1912.

[8]Harlan H. Barrows to Isaiah Bowman, October 6, 1923.

[9]Undated handwritten note on card by Isaiah Bowman, attached to letter, Harlan H. Barrows to Isaiah Bowman, October 6, 1923.

[10]Minutes of the Association of American Geographers, Notebook, 1904, p. 32.

[11]Geoffrey J. Martin, "On Association History, 1904-1954," unpublished paper, 1966.

[12]Geoffrey J. Martin, "The Ontographic Departure from Davisian Physiography," paper presented at the annual meeting of the Association of American Geographers, New York City, April 1976.

[13]Geoffrey J. Martin, "The Life and Thought of Isaiah Bowman," unpublished book-length manuscript.

[14]Grove Karl Gilbert to Nevin M. Fenneman, January 15, 1908.

[15]Henry G. Bryant to Albert Perry Brigham, January 24, 1913.

Henry G. Bryant to Isaiah Bowman, January 9, 1914.

[16]See the *Annals of the Association of American Geographers,* Vols. 1 to 14.

[17]"Genetic Geography: The Development of the Geographic Sense and Concept," *Annals of the Association of American Geographers,* Vol. 10, (1920), pp. 3-16.

[18]William Morris Davis to Albert Perry Brigham, October 12, 1909.

[19]Minutes of the Association of American Geographers, Notebook, 1904, p. 8.

[20]Minutes of the Association of American Geographers, Notebook, 1906, p. 8.

[21]Minutes of the Association of American Geographers, Notebook, 1906, p. 32.

[22]Minutes of the Association of American Geographers, Notebook, 1908, p. 57.

[23]Minutes of the Association of American Geographers, Notebook, 1908, p. 58.

[24]Minutes of the Association of American Geographers, Notebook, 1910, p. 81.

[25]Minutes of the Association of American Geographers, business meeting, December 30, 1914, unnumbered pages.

Katheryne Thomas Whittemore, "Celebrating Seventy-Five Years of the Journal of Geography 1897-1972," *Journal of Geography,* Vol. 71, (1972), pp. 7-18.

[26]*Annals of the Association of American Geographers,* Vol. 6, (1916), pp. 3-18.

[27]Isaiah Bowman to Nevin Fenneman, Marius Campbell, Douglas W. Johnson, François E. Matthes, and Eliot Blackwelder, April 18, 1915.

[28]Wolfgang L.G. Joerg, "The Subdivision of North America into Natural Regions: A Preliminary Inquiry," *Annals of the Association of American Geographers,* Vol. 4, (1914), pp. 55-83.

[29]Nevin M. Fenneman, "Physiographic Boundaries within the United States," *Annals of the Association of American Geographers,* Vol. 4, (1914), pp. 84-134.

[30]Curtis F. Marbut to Isaiah Bowman, February 15, 1914.

[31]Curtis F. Marbut to Isaiah Bowman, April 17, 1914.

[32]The correspondence of Nevin M. Fenneman is deposited in the archives of the University of Cincinnati, Ohio.

[33]Minutes of the Association of American Geographers, 1905, p. 28.

[34]William Morris Davis to Albert Perry Brigham, January 2, 1906.

[35]Minutes of the Association of American Geographers, 1910, p. 77

[36]Nevin M. Fenneman to Albert Perry Brigham, February 17, 1911.

[37]Statement by William Morris Davis, December 29, 1910.

[38]Pledge by Mark Jefferson, December 31, 1910.

[39]Pledge by Alfred H. Brooks, December 30, 1910.

[40]William D. Pattison, "The Star of the AAG," *The Professional Geographer,* Vol. 12, No. 5, (1960), pp. 18-19.

[41]The Memorandum of the American Geographical Society was confirmed by the Association of American Geographers by letter of Henry G. Bryant, January 6, 1914.

[42]Henry G. Bryant to Archer M. Huntington, April 16, 1913.

[43]*The Geographical Review,* Vol. 12, No. 3, (1922), p. 486.

[44]Isaiah Bowman to Richard E. Dodge, May 26, 1922.

[45]Isaiah Bowman to Richard E. Dodge, August 3, 1922.

[46]Isaiah Bowman to J. Paul Goode, February 10, 1917.

[47]J. Paul Goode to Isaiah Bowman, November 22, 1917.

[48]Harold W. Fairbanks to Albert Perry Brigham, October 18, 1905.

[49]Isaiah Bowman to Harold W. Fairbanks, February 2, 1915.

[50]Albert Perry Brigham to Henry G. Bryant, March 6, 1914.

[51]William Morris Davis to Isaiah Bowman, January 26, 1914.

Chapter 5

The Search for Alternatives, 1924-1943

[1]Chauncy D. Harris, "The Department of Geography of the University of Chicago in the 1930s and 1940s," *Annals of the Association of American Geographers,* Vol. 69, No. 1, (March, 1979), in press.

[2]"War Services of Members of the Association of American Geographers," *Annals of the Association of American Geographers,* Vol. 9, (1919), pp. 53-70.

[3]*Annals of the Association of American Geographers,* Vol. 11, (1921), pp. 130-131.

[4]*Annals of the Association of American Geographers,* Vol. 11, (1921).

[5]*University of California Publications in Geography,* Vol. 2, No. 2, (1925), pp. 19-53.

[6]Carl O. Sauer to Richard Hartshorne, June 22, 1946.

[7]John B. Leighly to Richard Hartshorne, November 6, 1975.

[8]Preston E. James and E. Cotton Mather, "The Role of Periodic Field Conferences in the Development of Geographical Ideas in the United States," *The Geographical Review,* Vol. 67, No. 4, (1977), pp. 446-461.

[9]"Progress in the Field of Mapping of Detailed Geographic Interrelationships," *Annals of the Association of American Geographers,* Vol. 17, (1927), pp. 26-27.

Vernor C. Finch, "Montford: A Study in Landscape Types in Southwestern Wisconsin," *Geographical Society of Chicago Bulletin,* 1933.

[10]Paper titles and abstracts available in the *Annals of the Association of American Geographers.*

[11]"The California Raisin Industry—A Study in Geographic Interpretation," *Annals of the Association of American Geography,* Vol. 14, No. 2, (1924), pp. 49-108.

[12]*Annals of the Association of American Geographers,* Vol. 14, No. 1, (1924), pp. 36-37.

[13]"Sequent Occupance," *Annals of the Association of American Geographers,* Vol. 19, No. 1, (1929), pp. 162-165.

[14]"Aeroplane Mapping on Isle Royale," *Annals of the Association of American Geographers,* Vol. 22, No. 1, (1932), p. 69.

[15]Isaiah Bowman, *The Pioneer Fringe,* (1931).

[16]"Some Geographic Relations in Trinidad," *The Scottish Geographical Magazine,* Vol. 42, (1926), pp. 84-93.

[17]*Annals of the Association of American Geographers,* Vol. 19, No. 2, (1929), pp. 67-109: *The Geographical Review* Vol. 26, No. 3, (1936), pp. 439-453.

[18]For the 1933 meeting see: "Conventionalizing Geographic Investigation and Presentation," *Annals of the Association of American Geographers,* Vol. 24, No. 2, (1934), pp. 77-122.

[19]Charles C. Colby, "Changing Currents of Geographic Thought in America," *Annals of the Association of American Geographers,* Vol. 26, No. 1, (1936), p. 29.

[20]See: the programs of the annual meetings, with the archives of the Association of American Geographers; see also printed titles and abstracts in the *Annals of the Association of American Geographers.*

[21]"Annual Meetings: Place and Presidential Addresses: 1904-1956," *The Association of American Geographers Handbook - Directory,* (1956), pp. 24-25.

[22]"The Geography of American Geographers," was first presented at the Evanston meeting of the National Council of Geography Teachers, December 28, 1933. Published: *The Journal of Geography,* Vol. 33, No. 6, (1934), pp. 221-236.

[23]Geoffrey J. Martin, "The Life and Thought of Isaiah Bowman," unpublished book-length manuscript.

[24]Richard Hartshorne, "Notes Toward a Bibliobigraphy of the Nature of Geography," 1978, unpublished, pp. 7-8.

[25]Ibid., pp. 11-12.

[26]Ibid., p. 30.

[27]Minutes of the Association of American Geographers Council Meeting, December 30, 1924.

[28]"The Royalties Fund," was established in order to facilitate research by grants ranging from $100 to $500. *The Association of American Geographers, Handbook - Directory,* (1956), p. 11.

[29]Report of the Secretary to the Council of the Association of American Geographers, 1936.

[30]Report of the Secretary to the Council of the Association of American Geographers, December 27, 1928, p. 2.

[31]Four page statement submitted to the Council of the Association of American Geographers, December 29, 1931.

[32]Preston E. James, address to the Association of American Geographers, New Orleans, April, 1978.

[33]Report of the Secretary to the Council of the Association of American Geographers, December 27, 1937.

[34]Almon E. Parkins to Lawrence Martin and Richard Hartshorne, February 1, 1934.

Chapter 6

*The American Society for Professional Geographers, 1943-1948 and
The Association of American Geographers, 1944-1948: A Time of Change*

[1]Geoffrey J. Martin, "The Life and Thought of Isaiah Bowman," unpublished book-length manuscript.

[2]"Washington a Major Center of Geography - Federal Government Employs 217 Professional Geographers," Association of American Geographers unpublished mimeographed statement, 2 pp.

See also: "Lessons from the War-time Experience for Improving Graduate Training for Geographic Research, Report of the Committee on Training and Standards in the Geographic Profession, National Research Council," *Annals of the Association of American Geographers,* Vol. 36, (1946), pp. 195-213.

[3]"Informal Personal Statement on Types of Membership for the Association of American Geographers," (1946), p. 1.

[4]Ralph H. Brown to Otto E. Guthe, April 8, 1944.

[5]*Annals of the Association of American Geographers,* Vol. 33, (1943), pp. 135-161.

[6]*Annals of the Association of American Geographers,* Vol. 33, (1943), pp. 163-195.

[7]Clyde F. Kohn to Shannon McCune, January 30, 1941.

[8]Wallace W. Atwood to W. Van Royen, January 12, 1943.

[9]J. Russell Smith to W. Van Royen, January 26, 1943.

[10]Stephen B. Jones to W. Van Royen, January 7, 1943.

[11]F. Webster McBryde, "The A.S.P.G. Story," unpublished statement, 15 pp. August, 1978.

[12]*Bulletin of the American Society for Geographical Research,* Vol. 1, No. 1, (1943), p. 1.

[13]F. Webster McBryde, "The A.S.P.G. Story," revised statement, unpublished, August 1978.

[14]William Van Royen made a thorough study of the constitutions of learned societies, and was most influential in drawing the constitution for the A.S.P.G. In 1948, when merger of the A.S.P.G. was accomplished, much of the A.S.P.G. constitution was retained.

[15]Otto Guthe to Ralph H. Brown, August 10, 1943.

[16]Stephen B. Jones to Ralph H. Brown, September 10, 1943.

[17]Eugene Van Cleef to Ralph H. Brown, August 29, 1943.

[18]Ralph H. Brown to Eugene Van Cleef, September 3, 1943.

[19]John K. Wright to Ralph H. Brown, June 3, 1943.

[20]Charles F. Brooks to Hugh H. Bennett, September 4, 1943.

[21]J. Russell Smith to Charles F. Brooks, September 9, 1943.

[22]Richard Hartshorne, "Draft of Recommendations," twelve Washington members of the Association of American Geographers, January 20, 1945.

See also: *Association of American Geographers Newsletter,* (June 1945), p. 2.

[23]"To the members of the A.A.G. who have been interested in sponsoring the election of candidates for membership," unpublished statement, November 19, 1943, one page.

[24]"Highlights in the History of the American Society for Professional Geographers during Its Separate Existence, 1943-1948," *The Association of American Geographers Handbook-Directory,* (1956), pp. 4-6.

[25]F. Webster McBryde, "The A.S.P.G. Story," unpublished statement, August 1978, 15 pp.

[26]Ibid.

[27]Clarence B. Odell, chairman of the membership committee of the ASPG, used the distribution list of the July 1946 AAG *Newsletter* in the drive for new members of the ASPG. Chauncy D. Harris, Secretary of the AAG, had compiled a list of 403 individuals teaching college geography courses who were available to the ASPG. Harris also served during 1946-1948 as chairman of the credentials committee of the ASPG, which then had four classes of membership, distinguishing between professional members and other members. He and George F. Deasy were in frequent touch about membership matters and had a close working relationship. Thus though there were substantial differences between the two organizations there was also close co-operation on many matters.

[28]E. Willard Miller, "The Contribution of the American Society for Professional Geographers to the Geographic Profession," unpublished statement, 1978, p. 6.

[29]Ibid., p. 6.

[30]Ibid., p. 21.

[31]*Association of American Geographers Newsletter,* (June, 1947), p. 2.

[32]The mailing list included members of the AAG (261); members of the ASPG as of June 18, 1946 (333); members of the National Council of Geography Teachers with university or college addresses as selected from the full membership list published in the *Journal of Geography,* February, 1946 (94); members of the Association of Pacific Coast Geographers as of April, 1946 (47); and teachers of geography courses in colleges or universities (403) compiled by Chauncy D. Harris from college catalogues on file in the U.S. Office of Education, October, 1945, numbers in parentheses excluding names on all preceding lists, total distribution in North America 1138. Copies were also supplied to members of the Institute of British Geographers. Personal communication from Chauncy D. Harris, October 4, 1978.

[33]Association of American Geographers. Forty-third Annual Meeting, Saturday through Monday, December 28, 29, and 30, 1946. The Ohio State University and The Deshler-Wallick Hotel, Columbus, Ohio. Revised programs.

[34]*Association of American Geographers Newsletter,* (June, 1947), p. 15.

[35]Ibid.

[36]Annual Meetings of American Geographers. Association of American Geographers, forty-fourth annual meeting; National Council of Geography Teachers, meeting of thirty-fourth year; American Society for Professional Geographers, fourth national meeting, Saturday through Wednesday, December 27, 28, 30, and 31, 1947, The University of Virginia, Charlottesville, Virginia. Revised program.

[37]Program. 1948 Annual Meetings. Joint Meeting of the Association of American Geographers Forty-fifth Annual Meeting and the American Society for Professional Geographers Fifth National Meeting at the University of Wisconsin, Madison, Wisconsin, December 27, 28, 29, 30, 1948.

[38]*Association of American Geographers Newsletter,* (July, 1946), pp. 1-2; (October, 1946), p. 4; (June, 1947), pp. 3-5; *Joint Newsletter of the Association of*

American Geographers and the American Society for Professional Geographers, (March, 1948), pp. 2-6.

[39]*Association of American Geographers Newsletter,* (July, 1946), pp. 4-5, (June, 1947), pp. 46-47; *Joint Newsletter of the Association of American Geographers and the American Society for Professional Geographers,* (June, 1948), pp. 26-27, 25; (September, 1948), p. 22.

[40]Chauncy D. Harris, "Informal Personal Statement on Types of Membership for the Association of American Geographers," unpublished, (1946), 2 pp.

[41]E. Willard Miller, (ASPG Secretary), "Notes on an Informal Meeting on Relations Between the A.A.G. and the A.S.P.G. Held September 23, 1946, in Washington, D.C." unpublished, p. 2.

[42]E. Willard Miller, "A Short History of the American Society for Professional Geographers," *The Professional Geographer,* New Series, Vol. 2, No. 1, (1950), pp. 29-40.

[43]"Transitional Steps," Anonymous five page mimeographed statement in the archives of the Association of American Geographers.

[44]"The President Speaks," *The Professional Geographer,* New Series, Vol. 1, No. 1, (1949), pp. 17-22.

[45]F. Webster McBryde to Geoffrey J. Martin, September 1, 1978.

[46]E. Willard Miller, "The Contribution of the American Society for Professional Geographers to the Geographic Profession," paper read at the annual meeting of the Association of American Geographers, New Orleans, 1978, p. 28.

Chapter 7

The Reconstituted Association of American Geographers: 1949-1963

[1]William L. Garrison, *Future Knowledge,* Department of Geography, University of Iowa, (1972).

[2]Stephen S. Visher, "The President and Vice-President of the AAG (1904-1948): Where they received their college and graduate training," *The Professional Geographer,* New Series, Vol. 2, No. 1, (1950), pp. 41-46.

[3]This Committee consisted of John Rose, Chairman; Richard Hartshorne, Lester Trueblood, Otis Starkey, and Meredith Burrill.

[4]The Constitutional Review Committee under the Chairmanship of Leslie Hewes was discharged with thanks by the Council in February 1952, and a special committee consisting of Meredith F. Burrill, Preston E. James, and Louis O. Quam was set up to draft a series of amendments to the Constitution.

[5]The 1954 Council had authorized the appointment of a Standing Committee of three (later reduced to one) to review, on assignment by the Council, questions pertaining to the Constitution and By-Laws.

[6]The Honorary Presidents are listed in the Appendices.

[7]An Association Membership Committee was active during the years 1949-1963, and was instrumental in the growth of the Association.

[8]*The Professional Geographer,* New Series, Vol. 1, (1949).

[9]Louis Peltier, "Analysis of Membership of the Association of American Geographers," *The Professional Geographer,* Vol. 4, No. 4, (1952), pp. 2-3.

[10]The dues of the Associates were later raised to $10, in 1956, but students could request a remission of 50 percent.

[11]Clarence F. Jones, "President's Analysis of the State of the Association," *The Professional Geographer,* Vol. 9, No. 3, (May, 1957), p. 16.

[12]Papers presented at annual meetings from 1949-1954 had to be submitted in full, two or three months before the meetings, so that they might be screened, and, if accepted for presentation, sent to discussants. Abstracts had to accompany the papers. In 1955, there was limited censorship and full papers were not required in advance. Abstracts, however, had to be submitted more than four months in advance of meetings. Copies of papers were evidently not required again until 1962, when they were due three months in advance of the annual meeting at Miami Beach. They were required also in 1963, fully four and a half months before the September deadline. Papers read were commonly limited to 15 minutes, although the time varied, from year to year, from 10 to 20 minutes.

[13]Preston E. James, *The Professional Geographer,* Vol. 11, No. 5, (1959), pp. 6-8.

[14]In 1951, 29 papers were submitted for publication, 15 were accepted. In the three succeeding years the numbers received and accepted were: 1952, 31 received, 14 accepted; 1953, 44 received, 14 accepted; 1954, 50 received, 17 accepted. The Editorial Board functioned until the end of 1963 when it was discontinued by the Council on the recommendation of the Editor of the *Annals.*

[15]The second map supplement was entitled, "The Floor of the World Ocean," drawn by Richard Edes Harrison and published in 1961. It was funded by the Geography Branch of the Office of Naval Research. The third map supplement, "Landforms of Utah -Proportional Relief," was composed by Merrill Ridd and published in 1963.

[16]Throughout the 1949-1963 era, abstracts of papers presented at annual meetings were published in either the June or September issues of the *Annals.* The Publication Committee was asked in 1959 to investigate the possibility of eliminating this practice, and to consider instead the publication of selected papers in a *Proceedings.* In 1960, the Council approved such a publication, to be established as soon as possible.

[17]Special cartography issues of *The Professional Geographer* appeared in November, 1950; September, 1951; September, 1952; November, 1952; November, 1955. The September, 1958, issue included a special Cartography Section.

[18]Upon Whittlesey's death in 1956, Andrew H. Clark accepted the editorship of the Monograph Series. He was succeeded in 1961 by Thomas R. Smith.

[19]The NAS/NRC Committees were first appointed in 1948-49. As listed in 1952, the final committees were chaired by Hoyt Lemons, Derwent Whittlesey, Raymond E. Murphy, Clyde F. Kohn, Richard Hartshorne, Andrew H. Clark, Meredith F. Burrill, Charles M. Davis, Arthur H. Robinson, and Harold M. Mayer, Members at large included Clarence F. Jones and Preston E. James (co-editors), John K. Wright (consulting editor), J. Russell Whitaker, G. Donald Hudson, and Robert S. Platt. The final publication of the book was paid for from AAG funds.

[20]*The National Atlas* was finally published in 1970. It contained 765 reference and thematic maps depicting the distribution of the nation's natural features, resources, historical evolution, human activites and conditions, administrative subdivisions, and place in world affairs.

[21]Howard G. Roepke, "Report of the Committee on Mass Data Processing Techniques," *The Professional Geographer,* Vol. 13, No. 2, (1961), pp. 23-25.

[22]It might be noted that at the annual meeting of the Association in Miami Beach, 1962, members demonstrated a considerable interest in digital computers and their applications.

[23]AAG delegates to the AAAS during the period 1949-1963 included: John K. Rose (1948-1954); Lester E. Klimm (1954-1957); Alfred H. Meyer (1957-1959); Alden D. Cutshall (1959-1963); John M. Loeffler (1963-1965).

[24]During the period 1949-1963, AAG representatives on the National Research Council were Edward B. Espenshade, Jr., (1949-1950); Joseph Van Riper (1950-1953); Guy Harold Smith (1953-1956); Louis O. Quam (1956-1959); Charles M. Davis (1959-1962); and Arthur Robinson (1962-1965).

[25]Louis C. Peltier, "The Potential of Military Geography," *The Professional Geographer,* Vol. 13, No. 6, (1961), pp. 1-5.

Albert H. Jackman, "The Nature of Military Geography," *The Professional Geographer,* Vol. 14, No. 1, (1962), pp. 7-12.

[26]Brian J.L. Berry, Richard L. Morrill and Waldo Tobler, "Geographic Ordering of Information: New Opportunities," *The Professional Geographer,* Vol. 16, No. 4, (1964), pp. 39-43.

[27]*The Professional Geographer,* Vol. 6, No. 4, (1954), p. 12.

[28]PAIGH was created in 1928 at Habana, Cuba, with headquarters in Mexico City. It is an official organization of 22 American States, including Canada, but currently not the government of Cuba. Affiliation is through the Department of State.

[29]*The Professional Geographer,* Vol. 10, No. 1, 1958 pp. 8-13.

[30]Chauncy D. Harris, "Visit of Soviet Geographers to the United States," *The Professional Geographer,* Vol. 13, No. 6, (1961), pp. 46-47.

[31]In the United States, affiliation with the IGU is through the National Academy of Sciences, and rests with the USA National Commission of the IGU. This served until 1952 as a Committee of the National Research Council, but in 1952 was placed in International Relations, headed by Wallace W. Atwood, Jr.

[32]John Reith, et al., *The Nontraditional Jobs of Geographers,* Association of American Geographers, Washington, D.C., (1960).

[33]*The Professional Geographer,* Vol. 6, No. 3, (1954), pp. 29-35.
[34]Available positions had been publicly announced for the first time five years earlier in the July, 1953 issue of *The Professional Geographer.*
[35]*The Professional Geographer,* Vol. 1, No. 1, (1949). pp. 15-16.
[36]G. Donald Hudson, "Professional Training of the Membership of the Association of American Geographers," *Annals of the Association of American Geographers,* Vol. 41, No. 2, (1951), pp. 97-115; Carl O. Sauer, "The Education of a Geographer," *Annals of the Association of American Geographers,* Vol. 46, No. 3, (1956), pp. 287-299.

C. Warren Thornthwaite, "The Task Ahead," *Annals of the Association of American Geographers,* Vol. 51, No. 4, (1961), pp. 345-356.
[37]The project was to be carried out in five stages: Stage 1, the definition of basic ideas and skills by a working group, including high school and university teachers, and research scholars in geography; Stage 2, the circulation of a statement prepared by the working committee in Stage 1 to a much larger group for comment; Stage 3, the development of experimental units by high school teachers working in cooperation with research scholars in nearby universities; Stage 4, the detailed planning of the course; and Stage 5, the production of the course. William D. Pattison and Henry J. Warman served respectively as director and coordinator of the project during its first three stages.

Chapter 8
New Directions, 1963-1978

[1]Meredith F. Burrill, "Actions of the Association of American Geographers Council, East Lansing, August 28, 1961," *The Professional Geographer,* Vol. 13, No. 6, (1961), p. 31.
[2]Arch C. Gerlach, "President's Report," *The Professional Geographer,* Vol. 15, No. 1, (1963), p. 27.
[3]Arch C. Gerlach, "President's Report," *The Professional Geographer,* Vol. 15, No. 5, (1963), p. 27.
[4]*Association of American Geographers Newsletter,* Vol. 9, No. 7, (August-September, 1974), p. 5.
[5]Donald J. Patton, ed., *From Geographic Discipline to Inquiring Student,* Final Report on the High School Geography Project, Association of American Geographers, (1970), 55 pp.
[6]Arvin W. Hahn, "Report to the President," *The Professional Geographer,* Vol. 15, No. 6, (1963), p. 31.
[7]Arvin W. Hahn, "Progress Report: Membership Drive," *The Professional Geographer,* Vol. 16, No. 1, (1964), p. 24.
[8]Arvin W. Hahn, "Report to the President," *The Professional Geographer,* Vol. 16, No. 2, (1964), p. 33.

[9]Arvin W. Hahn, "Report to the President," *The Professional Geographer*, Vol. 16, No. 4, (1964), p. 31.

[10]Norton S. Ginsburg, "Council Meeting, 1964-65," *The Professional Geographer*, Vol. 16, No. 4, (1964) p. 35.

[11]Norton S. Ginsburg, "Report of the Executive Committee of the Association of American Geographers Council," *The Professional Geographer*, Vol. 17, No. 2, (1965). p. 36.

[12]Among gifts presented to the Association in 1965-66 was a deed for 94.75 acres of land in New Hope Township, Chatham County, North Carolina, given by Dr. J. Sullivan Gibson of the Department of Geography, University of North Carolina. See: "A Generous Gift to the Association," *The Professional Geographer*, Vol. 18, No. 1, (January, 1966), p. 31.

[13]Norton S. Ginsburg, "Executive Committee Meeting," *The Professional Geographer*, Vol. 13, No. 1, (1966), p. 31.

[14]The announcement appears in full in *The Professional Geographer*, Vol. 13, No. 3, (1966), p. 170.

[15]"Report on 'CONPASS'," *The Professional Geographer*, Vol. 18, No. 6, (1966), pp. 366-368.

[16]Robert W. Peplies, "Commission on Geographic Applications of Remote Sensing," *Association of American Geographers Newsletter*, Vol. 1, No. 1, (November, 1967), p. 2.

[17]For a complete list of Commission on College Geography publications, see "The Final Report of the Commission on college Geography to the National Science Foundation," by John F. Lounsbury, 1974.

[18]Nan Lin, William D. Garvey, and Carnot E. Nelson, "A Study of the Communication Structure of Science," Report No. 10; "Some Comparisons of Communication Activities in the Physical and Social Sciences," Report No. 11, Johns Hopkins University, (1970).

[19]Richard E. Lonsdale, "Publications Committee Report," *The Professional Geographer*, Vol. 27, No. 3, (1975), p. 405.

[20]John S. Adams, "Council Meeting," *Association of American Geographers Newsletter*, Vol. II, No. 6, (1976), p. 14.

[21]Salvatore J. Natoli to Harm J. de Blij, April 21, 1978.

[22]*Association of American Geographers Newsletter*, Vol. 3, No. 8, (October, 1969), p. 2.

Wesley Calef, *The Professional Geographer*, Vol. 21, No. 6, (1969), p. 414.

[23]John R. Borchert, "1969 Association of American Geographers Meeting: Letter from the President," *Association of American Geographers Newsletter*, Vol. 2, No. 9, (1968), p. 1.

[24]*Association of American Geographers Newsletter*, Vol. 3, No. 10, (1969), p. 1.

[25]*The Professional Geographer*, Vol. 21, No. 4, (1969), p. 276.

[26]J. Warren Nystrom, "Annual Business Meeting," *The Professional Geographer*, Vol. 24, No. 1, (1972), p. 38.

[27]Harm J. de Blij, "Annual Business Meeting, Milwaukee," *The Professional Geographer*, Vol. 27, No. 3, (1975), p. 414.

[28]*The Professional Geographer,* Vol. 25, No. 3, (1973), p. 326.

[29]*Association of American Geographers Newsletter,* Vol. 11, No. 2, (1976), p. 1.

[30]Marvin W. Mikesell, "From the President," *Association of American Geographers Newsletter,* Vol. 10, No. 8, (1975), p. 1.

[31]"The Cost of Non-Registrants at Annual Meetings," *Association of American Geographers Newsletter,* Vol. 10, No. 10, (December, 1975), p. 1.

[32]Wesley Calef, "Summary of Council Minutes," *The Professional Geographer,* Vol. 22, No. 4, (1970), p. 227.

[33]Wesley Calef, "Minutes of the Annual Business Meetings," *The Professional Geographer,* Vol. 23, No. 1, (1971), p. 50.

[34]"Task Forces Outline Action Program for Association of American Geographers," *Association of American GeographersbNewsletter,* Vol. 5, No. 1, (1971), p. 1.

[35]*Association of American Geographers Newsletter,* Vol. 5, No. 7, (1971), p. 1.

[36]"Two Association of American Geographers Planning Development Task Force Proposals Funded," *Association of American Geographers Newsletter,* Vol. 6, No. 4, (1972), p. 1.

[37]"Geographers Take Their Place," *Mosaic,* Vol. 8, No. 2, (1977), pp. 38-44.

[38]"Project Development Committee Welcomes Submissions," *Association of American Geographers Newsletter,* Vol. 8, No. 9, (1973), p. 1.

"From the President," *Association of American Geographers Newsletter,* Vol. 11, No. 1, (1976), p. 3.

[39]John S. Adams, "Council Meeting," *Association of American Geographers Newsletter,* Vol. 12, No. 7, (1977), p. 18.

[40]D. Morgan and John S. Adams, "Results of the Membership Survey," (unpublished report), February, 1978, p. 3.

[41]Harm J. de Blij, "Council Meeting, Kansas City," *The Professional Geographer,* Vol. 24, No. 4, (1972), p. 346.

[42]*Association of American Geographers Newsletter,* Vol. 7, No. 10, (December, 1972), p. 1; Vol. 8, No. 1, (1973), pp. 1 and 2; Vol. 8, No. 5, (1973).

[43]"Results of Vote on Changes in Constitution and Bylaws and National Election," *The Professional Geographer,* Vol. 25, No. 3, (1973), p. 311.

[44]John S. Adams, "Report of Long Range Planning Committee," (unpublished), February, 1978, p. 31.

[45]"New Committee on Geographic Education," *Association of American Geographers Newsletter,* Vol. 3, No. 9, (1969), p. 1.

[46]Wesley Calef, "Council Meeting, San Francisco," *The Professional Geographer,* Vol. 23, No. 1, (1971), p. 46.

At the same Council meeting, John R. Borchert reported that a referendum on the merger question had taken place: 75 percent of members solely of the AAG or the NCGE supported the idea, while geographers who were members of both organizations divided 50 - 50 on the issue.

[47]Wesley Calef, "Council Meeting, Kansas City," *The Professional Geographer,* Vol. 24, No. 4, (1972), p. 341.

[48]Harm J. de Blij, "Council Meeting, Chicago," *The Professional Geographer,* Vol. 25, No. 2, (1973), p. 146.

[49]Wesley Calef, "Council Meeting, Kansas City," *The Professional Geographer,* Vol. 24, No. 4, (1972), p. 341.

[50]"Abstract Report of the Publications Committee on Serial Publications Policy," *The Professional Geographer,* Vol. 16, No. 2, (1964), pp. 39-43.

[51]J. Warren Nystrom, "First Council Meeting, 1968," *The Professional Geographer,* Vol. 20, No. 6, (1968), p. 415.

[52]Harm J. de Blij, "Council Meeting, Chicago," *The Professional Geographer,* Vol. 25, No. 2, (1973), p. 147.

[53]John S. Adams, "Council Meeting, New York," *Association of American Geographers Newsletter,* Vol. 11, No. 6, (1976), p. 13.

[54]*The Professional Geographer,* Vol. 16, No. 6, (1964). pp. 25-27.

[55]At this time the name is subject to change. The first number is anticipated in 1979.

[56]Written by Brian J.L. Berry pursuant to decisions made at the Association's Council meeting, Milwaukee, Wisconsin, October 21, 1978.

Chapter 9

The Association of American Geographers; Retrospect

[1]Norton Ginsburg, *"The Mission of a Scholarly Society,"* The Professional Geographer, Vol. 24, No. 1, (February, 1972), p. 2.

[2]The Constitution of the Association of American Geographers, 1904, p. 1.

[3]William Morris Davis to Isaiah Bowman, March 18, 1906.

[4]Charles C. Colby to Geoffrey J. Martin, undated.

[5]Marcel Aurousseau to Geoffrey J. Martin, January 20, 1961.

[6]J. Russell Smith to Geoffrey J. Martin, undated but 1962.

[7]This format was retained until March, 1923, when *"Supplement to Annals of the Association of American Geographers"* made its first appearance. A feature of this *"Supplement,"* was a list of recent publications by each Association member. By 1935 the publication had required a new title, *"News Letter: For Circulation Among the Membership of the Association of American Geographers,"* and in May, 1936, the title *"Association of American Geographers: News Letter"* was adopted. This was revised to *"Newsletter of the Association of American Geographers"* in July, 1946. Notwithstanding these changes of name, the *"Newsletter"* discharged a vital function for the membership; it was cheap, quickly produced, and easily mailed.

[8]L.P. Denoyer to Albert Perry Brigham, October 22, 1912.

[9]Geoffrey J. Martin, *Mark Jefferson: Geographer,* (1968), "Map Making: The Inquiry and Paris Peace Conference," chapter 8, pp. 167-198.

219

Notes

[10]Carl Sauer in Maynard Weston Dow's "Geographers on Film" series. Film number one.

[11]Preston E. James and E. Cotton Mather, "The Role of Periodic Field Conferences in the Development of Geographical Ideas in the United States," *The Geographical Review,* Vol. 67, No. 4 (1977), pp. 446-461.

[12]John K. Wright, "What's 'American' About American Geography?" *Human Nature in Geography,* (1966), chapter 8, pp. 124-139.

[13]Geoffrey J. Martin, "On Association History, 1904-1954," unpublished paper, 1966.

[14]Richard E. Dodge to Albert Perry Brigham, February 2, 1913.

APPENDIX A

PROPOSED CONSTITUTION
OF THE
AMERICAN GEOGRAPHERS ASSOCIATION
(1904)

I. Name and Object.

The name of this organization shall be the American Geographers Association. Its object shall be the cultivation of the scientific study of geography in all its branches, especially by promoting acquaintance, intercourse and discussion among its members, by encouraging and aiding geographical exploration and research, by assisting the publication of geographical essays, by developing better conditions for the study of geography in schools, colleges, and universities, and by cooperating with other societies in the development of an intelligent interest in geography among the people of North America.

II. Officers.

The officers of the Association shall be a President, two Vice-Presidents, a Secretary, a Treasurer (one person may be both secretary and treasurer), and two Councillors. These officers shall constitute a Council which shall manage the affairs of the Association. Nominations of officers, made by a committee of three members previously appointed for that purpose, must be sent to members not less than sixty days before the annual meeting of the Association. Any five members may make independent nominations, which if received by the Secretary thirty-five days before the annual meeting, must be sent to all members not less than thirty days before the meeting.

The election of officers must be by ballot. Each member may make such changes as he wishes in the ballot received from the

Secretary. He should then enclose the ballot in a sealed envelope, sign his name on the outside, and send or give it to the Secretary in time for it to be counted at the annual meeting.

III. Membership.

Membership shall be limited to persons who have done original work in some branch of geography. A nomination for membership must be made on an offical blank, prepared by the Council, signed by two members of the Association, and sent to the Secretary. The Council shall consider all nominations: if its action is favorable, the nominee will be recommended to the Association for election: if unfavorable, the nomination will not be brought before the Association, except on the written request of both signers of the nomination and with a statement of the Council's action. All elections to membership must be by ballot, at such times as the Council may determine. Members resident in North America shall pay an annual fee of $5.00; but the Council may remit this fee, *sub silencio.* The names of members two years in arrears shall be stricken from the list of the Association.

IV. Meetings.

The annual meeting of the Association shall be held within a week of January 1, and ordinarily in connection with the meeting of the American Association for the Advancement of Science: but the Council may arrange other meetings in place of or in addition to this annual meeting. Announcement of the time and place of all meetings must be mailed to members at least 30 days in advance.

V.

Changes in Constitution or By-Laws may be made by two-thirds vote of the members present at any meeting, provided that printed notice of the proposed change has been sent to all members with the call for the meeting, and that at least fifteen members are present.

BY-LAWS

1. No regular publication will for the present be issued by the Association.

2. Members are free to publish, in any way they desire, essays that have been submitted to the Association.

3. Reports of meetings of the Association, prepared under supervision of the Secretary, shall be sent for publication to such journals as the Council may direct.

CONSTITUTION AND BYLAWS OF THE ASSOCIATION OF AMERICAN GEOGRAPHERS (In effect, 1978)

Article I—Name

The name of the organization shall be the Association of American Geographers.

Article II—Objectives

The objectives of this Association shall be to further professional investigations in geography and to encourage the application of geographic findings in education, government, and business. The Association shall support these objectives by promoting acquaintance and discussion among its members and with scholars in related fields, by stimulating research and scientific exploration, by encouraging the publication of scholarly studies, and by performing services to aid the advancement of its members and the field of geography.

Article III—Membership

SECTION 1. Individual Members. Persons who are interested in the objectives of the Association are eligible for membership.

SECTION 2. Institutional Members. Corporations, firms, institutions, libraries, departments, and other scientific, education, and/or business associations interested in the objectives of the Association may on invitation of or application to the Council become

Institutional Members. The Council at its discretion shall determine the types, classes, or categories of such membership.

SECTION 3. Membership Rights. Members in good standing shall have full rights to nominate candidates for national and their respective regional offices, vote thereon, and hold such offices if duly elected; they shall be entitled to participate, under applicable rules in meetings, programs, and other activities and services of the National Organization and their respective regional subdivisions.

Article IV—Officers, Council, and Committees

SECTION 1. Officers, Councillors, and Elected Committees. The elected officers of the Association shall be a President, a Vice-President, a Secretary, and a Treasurer. The Councillors shall be six elected at large, and one elected from and by each Regional Division. The duties of the President, Vice-President and Treasurer shall be those normally pertaining to their posts. The Secretary shall serve as a Secretary of the Council and the Executive Committee. A Nominating Committee and an Honors Committee shall be elected annually. Terms of office shall begin on July 1 following the Annual Meeting of the Association and the period between Annual Meetings shall be considered a one-year term. The terms of office shall be one year for President, Vice-President, and members of the Nominating Committee and Honors Committee, three years for Secretary, Treasurer and Councillors. The President, Vice-President, and Councillors shall not be eligible for immediate reelection to the same office, and the Secretary and Treasurer shall not be eligible for reelection to the same office until after a lapse of six years following termination for the first tenure. The terms of office of the Councillors elected at large shall be arranged so that two shall retire each year. In addition to Councillors elected at large, each Regional Division shall elect by mail ballot one Councillor from that region, also for a three-year term.

SECTION 2. An Executive Director appointed by the Council shall maintain the Central Office of the Association, and perform such other duties as the Council may direct.

SECTION 3. Methods of Nomination and Election of Officers. The Nominating Committee shall make two or more nominations for each office, except that the Vice-President may be named as a single candidate for the Presidency. However, if the Vice-President is not in a position to accept candidacy, the Nominating Committee must nominate at least two candidates for the Presidency. The Nominating

Committee shall submit its slate of candidates to the Council at least five months prior to the next Annual Meeting of the Association. The membership shall be promptly notified of these nominations. Additional nominations may be made in writing by any 25 members of the Association if received in the Central Office at least 90 days prior to the announced date of the Annual Meeting. At least 45 days before the Annual Meeting official ballots shall be mailed to all Members, to be returned within 30 days and not be counted by tellers from the list of Members. The Council shall have power to fill vacancies until the next election.

SECTION 4. Council and Executive Committee. The Council shall consist of the officers and councillors elected under Section I, the most recent Past President, and ex officio, the Executive Director. The Council shall have power to transact all business of the Association, establish committees, and assign specific responsibilities to the various officers and committees of the Association. The Council may delegate to officers and to the Executive Director authority to sign contracts. The Council shall meet at least once each year at the call of the President. Notices of Council meetings shall be sent out at least two weeks in advance. A majority of the Council shall constitute a quorum.

The Executive Committee shall be the President, Vice-President, Secretary, Treasurer and the most recent Past President. The Executive Committee may invite other members to participate in discussion of matters within their special competence. The Executive Committee shall meet when necessary to transact the business of the Association. Official actions of the Executive Committee shall be subject to approval by the Council by majority vote in a mail ballot. Official actions of the Executive Committee and by the Council shall be published as promptly as practicable.

SECTION 5. Editors and Assistant Editors shall be appointed by the Council for such terms as the Council may determine.

SECTION 6. Committees. A Nominating Committee of three Members and an Honors Committee of three Members shall be elected at the Business Meeting of the Association. The Council shall make at least three nominations each for members of the Nominating Committee and of the Honors Committee. The nominations by the Council shall be posted, along with the agenda for the Business Meeting, during the opening day of the Annual Meeting. For these Committees, additional candidates may be nominated, either in advance by mail, if supported by at least 10 Members, or at the Business Meeting, if seconded. Two years must elapse before a Past President can be a member of the Nominating Committee. Members of other committees may be appointed by the President, subject to the approval of the Council, and shall act according to procedures established by the Council.

Article V—Meetings

SECTION 1. Annual Meetings. The Annual Meeting of the Association shall be held at such time and place as the Council may designate. The Council may arrange other meetings in addition to the Annual Meeting. Announcement of the time and place of meetings must be mailed to Members at least 30 days in advance.

SECTION 2. Business Meetings. A Business Meeting shall be held during the Annual Meeting. During the Business Meeting there shall be reports of the officers, election of the elected committees, and such other business as has been placed on the Agenda by the Council or has been proposed by the membership under pertinent rules established by the Council within the scope of Article IV, Section 4, of this constitution.

SECTION 3. All resolutions adopted by the Council or by an Annual Business Meeting must fall within the scope of the objectives of the Association of American Geographers as stated in Article II of the Constitution; those outside the scope of these objectives are to be ruled out of order. Resolutions must be posted conspicuously by the Secretary at least 24 hours before the Annual Business Meeting and be distributed at the meeting.

Article VI—Regional Divisions

SECTION 1. Establishment of Regional Divisions. The Association by vote of the Council may establish Regional Divisions in specific areas and may contribute toward the operation of these Divisions. Such Divisions shall promote the objectives of the Association in their respective areas. Upon the establishment of a Division, a Chairman and a Secretary-Treasurer shall be appointed by the Council. After an initial term of the appointed officers not to exceed two years, all officers shall be elected by the Members of the Division. The Council shall determine the boundaries of the Division. A Division may be disbanded for inactivity or other cause, such disbanding to be on recommendation of the Council by majority vote of Members voting at the Business Meeting of the Association.

SECTION 2. Officers and Duties. Each Division shall have a Chairman and such other officers and committees as the Division may authorize. All officers and the regional Councillors shall be members of the Association. The Chairman shall serve for not more than two consecutive years.

SECTION 3. Local Chapters. Subject to approval by the Council, the Divisions may authorize local chapters.

Article VII—Changes in the Constitution

Changes in the Constitution proposed either by the Council or by petition of 100 Members may be made by affirmative vote of majority of Members voting in either of two ways: first, at any regular meeting by ballot mailed or handed to the Secretary, provided that printed notice of the proposed change was mailed to all Members with the call of the meeting; second, by mail ballot at any time, provided that 60 days notice of the proposed change has been mailed to all Members.

BYLAWS

SECTION 1. Dues. All members shall pay an annual fee as set by the Council and ratified at the Annual Business Meeting, or by mail vote. The Council may waive this in those individual cases that warrant special consideration. Members may obtain a waiver of further payment of the annual fee by making a single payment equal to twenty times the current annual fee; payments thus made shall be invested in the name of the Association and the income from such investment shall be regarded as dues. Institutional Members shall pay an annual fee determined by the Council as appropriate for the type or class represented. Members in arrears shall be dropped from the Association after due notice, according to the procedures established and announced by the Council. A Member in good standing at the time of resignation may be reinstated on application. Members dropped for non-payment of dues may be reinstated on payment of that portion of the year in arrears during which they received unpaid-for publications of the Association.

SECTION 2. Honors. The Association shall encourage meritorious achievements in geography by awarding honors in special recognition of outstanding contributions toward the advancement or welfare of the profession. The contributions recognized might be in research, applied research, writing, teaching, committee work, administrative work, collaborative work with nongeographers, or in other aspects of geographic professional work. Recognition may take the form of grants for further advancing or implementing the contributions already made, citations, medals, an annual honor roll or plaque, or other form appropriate to the purpose of such recognition. The Honors Committee shall submit to the Council nominations for such awards accompanied by a statement indicating the contribution which forms the basis of the proposed award. At the Annual Meeting the Council shall announce the award of such honors as it may have approved.

SECTION 3. It shall be the responsibility of the Past President to address the Annual Meeting.

SECTION 4. Research and Publication Funds. The Association shall receive and administer funds in support of research and publication in the field of geography.

A Research Grants Committee of three Members shall consider applications for research grants and recommend to the Council the granting of appropriate sums in view of the projects submitted and funds available. Proposals for research grants shall be submitted through the Central Office of the Association.

A Publications Committee consisting of a member of the Council as Chairman, the editors and two additional Members of the Association, shall advise the Council on publication policy and the funding thereof.

SECTION 5. Petition and mail vote. The Council will consider a petition by any member or group of members to initiate an action, or reconsider a previous action, taken by the Council or an Annual Business Meeting. Matters of concern to the Association may be submitted by the Council to the Membership for a mail vote at any time.

SECTION 6. Publications. The Association shall issue such publications as the Council may determine.

SECTION 7. Signatures. The Council of the Association has sole authority to designate persons eligible to issue checks, sign other financial documents, or otherwise represent the Association as its agent.

SECTION 8. Amendments, The Bylaws may be amended by a majority of the Members voting at the Business Meeting of the Association, such vote to be followed by, and to take effect upon, ratification by a mail vote if the Council shall so determine.

SECTION 9. In the event that 50 members of the Association consider that a resolution adopted by Council or by the Annual Business Meeting is unrelated to the objectives set forth in Article II of the Constitution, the officers of the Association are required to submit the question of its appropriateness to a mail vote of all members of the Association.

APPENDIX B

OFFICERS OF THE ASSOCIATION OF AMERICAN GEOGRAPHERS

1904 – Pres., William M. Davis; Sec., Albert Perry Brigham.

1905 – Pres., William M. Davis; Vice-Presidents, Grove K. Gilbert and Angelo Heilprin; Sec., Albert P. Brigham; Treas., Albert P. Brigham; Councilors, Cyrus C. Adams, Henry C. Cowles, and Ralph S. Tarr.

1906 – Pres., Cyrus C. Adams; 1st Vice-Pres., Angelo Heilprin, 2nd Vice-Pres., William Libbey; Sec., Albert P. Brigham; Treas., Albert P. Brigham; Councilors, Henry C. Cowles, William M. Davis, and Israel C. Russell.

1907 – Pres., Angelo Heilprin; 1st Vice-Pres., Ralph S. Tarr, 2nd Vice-Pres., G.W. Littlehales; Sec., Albert P. Brigham; Treas., Albert P. Brigham; Councilors, Cyrus C. Adams, William M. Davis, and John P. Goode.

1908 – Pres., Grove K. Gilbert; 1st Vice-Pres., Rollin D Salisbury; 2nd Vice-Pres., Ellen C. Semple; Sec. Albert P. Brigham; Treas., Nevin M. Fenneman; Councilors, Cyrus C. Adams, William M. Davis, and Ralph S. Tarr.

1909 – Pres., William M. Davis; 1st Vice-Pres., Louis A. Bauer; 2nd Vice-Pres., Emory R. Johnson; Sec., Albert P. Brigham; Treas. Nevin M. Fenneman; Councilors, Cyrus C. Adams, Richard E. Dodge, and Ralph S. Tarr.

1910 – Pres., Henry C. Cowles; 1st Vice-Pres., Henry Gannett; 2nd Vice-Pres., Mark Jefferson; Sec. Albert P. Brigham; Treas., Nevin M. Fenneman; Councilors, William M. Davis, Richard E. Dodge, and Ralph S. Tarr.

1911 – Pres., Ralph S. Tarr; 1st Vice-Pres., Alfred H. Brooks; 2nd Vice-Pres., Henry G. Bryant; Sec., Albert P. Brigham; Treas., Nevin M. Fenneman; Councilors, William M. Davis, Richard E. Dodge, and Herbert E. Gregory.

1912 – Pres., Rollin D Salisbury; 1st Vice-Pres., Marius R. Campbell; 2nd Vice-Pres., Isaiah Bowman; Sec., Albert P. Brigham; Treas., Nevin M. Fenneman; Councilors, William M. Davis, Herbert E. Gregory, and Lawrence Martin.

1913 – Pres., Henry G. Bryant; 1st Vice-Pres., Ellsworth Huntington; 2nd Vice-Pres., Charles C. Adams; Sec., Albert P. Brigham; Treas., François E. Matthes; Councilors, Herbert E. Gregory, Lawrence Martin, and Robert DeC. Ward.

1914 – Pres., Albert P. Brigham; 1st Vice-Pres., Curtis F. Marbut; 2nd Vice-Pres., Charles R. Dryer; Sec., Isaiah Bowman; Treas., François E. Matthes; Councilors, Alfred H. Brooks, Lawrence Martin, and Robert DeC. Ward.

1915 – Pres., Richard E. Dodge; 1st Vice-Pres., Mark Jefferson; 2nd Vice-Pres., Frank Carney; Sec., Isaiah Bowman; Treas., François E. Matthes; Councilors, Alfred H. Brooks, William Libbey, and Robert DeC. Ward.

1916 – Pres., Mark Jefferson; 1st Vice-Pres., J. Russell Smith; 2nd Vice-Pres., J. Paul Goode; Sec., Isaiah Bowman; Treas., François E. Matthes; Councilors, Alfred H. Brooks, William Libbey, and Ray H. Whitbeck.

1917 – Pres., Robert DeC. Ward; 1st Vice-Pres., Harlan H. Barrows; 2nd Vice-Pres., William H. Hobbs; Sec., Richard E. Dodge; Treas. François E. Matthes; Councilors, Albert P. Brigham, William Libbey, and Ray H. Whitbeck.

1918 – Pres., Nevin M. Fenneman; 1st Vice-Pres., Charles R. Dryer; 2nd Vice-Pres., Bailey Willis; Sec., Oliver L. Fassig; Treas., François E. Matthes; Councilors, Albert P. Brigham, Walter S. Tower, and Ray H. Whitbeck.

1919 – Pres., Charles S. Dryer; 1st Vice-Pres., Herbert E. Gregory; 2nd Vice-Pres., Isaiah Bowman; Sec., Oliver L. Fassig; Treas., François E. Matthes; Councilors, Eliot Blackwelder, Albert P. Brigham, and Walter S. Tower.

1920 – Pres., Herbert E. Gregory; 1st Vice-Pres., Harlan H. Barrows; 2nd Vice-Pres., Charles F. Brooks; Sec., Richard E. Dodge;

Treas., George B. Roorbach; Councilors, Eliot Blackwelder, Walter S. Tower, and Ray H. Whitbeck.

1921 – Pres., Ellen C. Semple; 1st Vice-Pres., Alfred J. Henry; 2nd Vice-Pres., Curtis F. Marbut; Sec., Richard E. Dodge; Treas., George B. Roorbach; Councilors, Eliot Blackwelder, Charles R. Dryer, Nevin M. Fenneman, Herbert E. Gregory, and Ray H. Whitbeck.

1922 – Pres., Harlan H. Barrows; Vice-Pres., Alfred H. Brooks; Sec., Richard E. Dodge; Treas., George B. Roorbach; Councilors, Nevin M. Fenneman, Herbert E. Gregory, W.L.G. Joerg, Ellen C. Semple, and Ray H. Whitbeck.

1923 – Pres., Ellsworth Huntington; Vice-Pres., Nels A. Bengtson; Sec., Charles C. Colby; Treas. George B. Roorbach; Councilors, Harlan H. Barrows, Nevin M. Fenneman, W.L.G. Joerg, George R. Mansfield, and Ellen C. Semple.

1924 – Pres., Curtis F. Marbut; Vice-Pres., Oliver E. Baker; Sec., Charles C. Colby, Treas., Vernor C. Finch; Councilors, Harlan H. Barrows, Ellsworth Huntington, W.L.G. Joerg, Wellington D. Jones, and George R. Mansfield.

1925 – Pres., Ray H. Whitbeck; Vice-Pres., Homer L. Shantz; Sec., Charles C. Colby; Treas., Vernor C. Finch; Councilors, Oliver E. Baker, Ellsworth Huntington, Wellington D. Jones, George R. Mansfield, and Curtis F. Marbut.

1926 – Pres., J. Paul Goode; Vice-pres., George B. Roorbach; Sec., Charles C. Colby; Treas., Vernor C. Finch; Councilors, Oliver E. Baker, Wellington D. Jones, Curtis F. Marbut, Philip S. Smith, and Ray H. Whitbeck.

1927 – Pres., Marius R. Campbell; Vice-Pres., Charles C. Adams; Sec., Charles C. Colby; Treas., Vernor C. Finch; Councilors, Oliver E. Baker, Darrell H. Davis, J. Paul Godoe, Philip S. Smith, and Ray H. Whitbeck.

1928 – Pres., Douglas W. Johnson; Vice-Pres., W.L.G. Joerg; Sec., Charles C. Colby; Treas., Vernor C. Finch; Councilors, Marius R. Campbell, Darrell H. Davis, J. Paul Goode, Kenneth C. McMurry, Philip S. Smith, and Ray H. Whitbeck.

1929 – Pres., Lawrence Martin; Vice-Pres., Robert M. Brown; Sec., Darrell H. Davis; Treas., Robert S. Platt; Councilors, Marius R. Campbell, Mark Jefferson, Douglas W. Johnson, and Kenneth C. McMurray.

1930 – Pres., Almon E. Parkins; Vice-Pres., Frank E. Williams; Sec., Darrell H. Davis; Treas., Robert S. Platt; Councilors, Mark Jefferson, Douglas W. Johnson, Lawrence Martin, Kenneth C. McMurry, and John K. Wright.

1931 – Pres., Isaiah Bowman; Vice-Pres., George D. Hubbard; Sec., Darrell H. Davis; Treas., Robert S. Platt; Councilors, Mark Jefferson, Lawrence Martin, Almon E. Parkins, Glenn T. Trewartha, and John K. Wright.

1932 – Pres., Oliver E. Baker; Vice-Pres., William H. Haas; Sec., Frank E. Williams; Treas., Robert S. Platt; Councilors, Isaiah Bowman, Almon E. Parkins, Glenn T. Trewartha, Derwent Whittlesey, and John K. Wright.

1933 – Pres., François E. Matthes; Vice-Pres., Stephen S. Visher; Sec., Frank E. Williams; Treas., Robert S. Platt; Councilors, Oliver E. Baker, Isaiah Bowman, Ralph H. Brown, Glenn T. Trewartha, and Derwent Whittlesey.

1934 – Pres., Wallace W. Atwood; Vice-Pres., Vernor C. Finch; Sec., Frank E. Williams; Treas. Robert S. Platt; Councilors, Oliver E. Baker, Ralph H. Brown, Preston E. James, François Matthes, Derwent S. Whittlesey.

1935 – Pres., Charles C. Colby; Vice-Pres., Claude H. Birdseye; Sec., Frank E. Williams; Treas., John E. Orchard; Councilors, Wallace W. Atwood, Ralph H. Brown, Kirk Bryan, Preston E. James, and François E. Matthes.

1936 – Pres., William H. Hobbs; Vice-Pres., John K. Wright; Sec., Preston E. James; Treas., John E. Orchard; Councilors, Wallace W. Atwood, Claude H. Birdseye, Kirk Bryan, Charles C. Colby, and Richard J. Russell.

1937 – Pres., W.L.G. Joerg; Vice-Pres., Guy-Harold Smith; Sec., Preston E. James; Treas., John E. Orchard; Councilors, Kirk Bryan, Charles C. Colby, Richard Harthorne, William H. Hobbs, and Richard J. Russell.

1938 – Pres., Vernor C. Finch; Vice-Pres., Griffith Taylor; Sec., Preston E. James; Treas., Guy-Harold Smith; Councilors, Richard Harthorne, William H. Hobbs, W.L.G. Joerg, Clarence F. Jones, and Richard J. Russell.

1939 – Pres., Claude H. Birdseye; Vice-Pres., George McC. McBride; Sec., Preston E. James; Treas., Guy-Harold Smith; Councilors, Vernor C. Finch, Richard Hartshorne, W.L.G. Joerg, Clarence F. Jones, and Lewis F. Thomas.

1940 – Pres., Carl O. Sauer; Vice-Pres., Forrest Shreve; Sec., Preston E. James; Treas., Guy-Harold Smith; Councilors, Claude H. Birdseye, Vernor C. Finch, Clarence F. Jones, Lewis F. Thomas, and J. Russell Whitaker.

1942 – Pres., J. Russell Smith; Vice-Pres., Nels A. Bengtson; Sec., Ralph H. Brown; Treas., Guy-Harold Smith; Councilors, S. Whittemore Boggs, Robert B. Hall, Carl O. Sauer, Griffith Taylor, and J. Russell Whitaker.

1943 – Pres., Hugh H. Bennett; Vice-Pres., Robert S. Platt; Sec., Ralph H. Brown; Treas., Guy-Harold Smith; Councilors, S. Whittemore Boggs, Edwin J. Foscue, Robert B. Hall, J. Russell Smith, and Griffith Taylor.

1944 – Pres., Derwent Whittlesey; Vice-Pres., Francis J. Marschner; Sec., Ralph H. Brown; Treas., Guy-Harold Smith; Councilors, Hugh H. Bennett, Edwin J. Foscue, Lester E. Klimm, J. Russell Smith, and J. Russell Whitaker.

1945 – Pres., Robert S. Platt; Vice-Pres., Vilhjalmur Stefansson; Sec., Ralph H. Brown; Treas., Guy-Harold Smith; Councilors, Hugh H. Bennett, Edwin J. Foscue, Lester E. Klimm, Philip S. Smith, and Derwent Whittlesey.

1946 – Pres., John K. Wright; Vice-Pres., John B. Leighly; Sec., Chauncy D. Harris; Treas., Guy-Harold Smith; Councilors, Stephen B. Jones, Lester E. Klimm, Robert S. Platt, Philip S. Smith, and Derwent Whittlesey.

1947 – Pres., Charles F. Brooks; Vice-Pres., Clarence F. Jones; Sec., Chauncy D. Harris, Treas., Guy-Harold Smith; Councilors, George B. Cressey, Stephen B. Jones, Robert S. Platt, Philip S. Smith, and John K. Wright.

1948 – Pres., Richard J. Russell; Vice-Pres., Clifford M. Zierer; Sec., Chauncy D. Harris; Treas., Charles B. Hitchcock; Councilors, Charles F. Brooks, George B. Cressey, Stephen B. Jones, John K. Rose, and John K. Wright.

1949 – Pres., Richard Hartshorne; Vice-Pres., Shannon McCune; Sec., Walter W. Ristow; Treas., Charles B. Hitchcock; Councilors, Wallace W. Atwood, Jr., George B. Cressey, George F. Deasy, Walter Kollmorgen, George J. Miller, John K. Rose, and John C. Weaver.

1950 – Pres., G. Donald Hudson; Vice-Pres., Raymond E. Murphy; Sec., Walter W. Ristow; Treas., Lloyd D. Black; Councilors, Wallace W. Atwood, Jr., Richard Hartshorne, George J. Miller, John K. Rose, Joseph A. Russell, Paul A. Siple, and John C. Weaver.

1951 – Pres., Preston E. James; Vice-Pres., Loyal Durand, Jr.; Sec., Louis O. Quam; Treas., Lloyd D. Black; Councilors, Wallace W. Atwood, Jr., Jan Broek, G. Donald Hudson, Raymond E. Murphy, Joseph A. Russell, Paul A. Siple, and John C. Weaver.

1952 – Pres., Glenn T. Trewartha; Vice-Pres., Samuel S. Van Valkenburg; Sec., Louis O. Quam; Treas., Lloyd D. Black; Councilors, Jan Broek, Robert M. Glendinning, Harold A. Hoffmeister, Preston E. James, Raymond E. Murphy, Joseph A. Russell, and Paul A. Siple.

1953 – Pres., J. Russell Whitaker; Vice-Pres., Joseph A. Russell; Sec., Louis O. Quam; Treas., Hoyt Lemons; Councilors, Jan Broek, Edward B. Espenshade, Jr., Robert M. Glendinning, Harold A. Hoffmeister, Shannon McCune, Raymond E. Murphy, and Glenn T. Trewartha.

1954 – Pres., Joseph A. Russell; Vice-Pres., Louis O. Quam; Sec., Burton W. Adkinson; Treas., Hoyt Lemons; Councilors, Edward B. Espenshade, Jr., Robert M. Glendinning, Leslie Hewes, Harold A. Hoffmeister, Trevor Lloyd, Shannon McCune, and J. Russell Whitaker.

1955 – Pres., Louis O. Quam; Vice-Pres., Clarence F. Jones; Sec., Burton W. Adkinson; Treas., Hoyt Lemons; Councilors, Edward A. Ackerman, Edward B. Espenshade, Jr., Wilma B. Fairchild, Leslie Hewes, Trevor Lloyd, Shannon McCune, and Joseph A. Russell.

1956 – Pres., Clarence F. Jones; Vice-pres., Chauncy D. Harris; Sec., Burton W. Adkinson; Treas., Wallace W. Atwood, Jr.; Councilors, Edward A. Ackerman, Wilma B. Fairchild, F. Kenneth Hare, Leslie Hewes, Trevor Lloyd, Richard F. Logan, and Louis O. Quam.

1957 – Pres., Chauncy D. Harris; Vice-Pres., Lester E. Klimm; Sec., Arch C. Gerlach; Treas., Wallace W. Atwood, Jr,; Councilors, Edward A. Ackerman, Loyal Durand, Jr., Wilma B.

Fairchild, F. Kenneth Hare, Henry M. Kendall, Richard F. Logan, Clarence F. Jones.

1958 – Pres., Lester E. Klimm; Vice-Pres., Paul A. Siple; Sec., Arch C. Gerlach; Treas., Wallace W. Atwood, Jr.; Councilors, Loyal Durand, Jr., F. Kenneth Hare, Chauncy D. Harris, Henry M. Kendall, Richard F. Logan, G. Etzel Pearcy, Kirk H. Stone.

1959 – Pres., Paul A. Siple; Vice-Pres., Jan O.M. Broek; Sec., Arch C. Gerlach; Treas., George F. Deasy; Councilors, Loyal Durand, Jr., Henry M. Kendall, Lester E. Klimm, Fred B. Kniffen, Alfred H. Meyer, G. Etzel Pearcy, Kirk H. Stone.

1960 – Pres., Jan O.M. Broek; Vice-Pres., Gilbert F. White; Sec., Meredith F. Burrill; Treas., George F. Deasy; Councilors, Fred B. Kniffen, Alfred H. Meyer, James J. Parsons, G. Etzel Pearcy, Paul A. Siple, Kirk H. Stone, Arthur H. Robinson.

1961 – Pres., Gilbert F. White; Vice-Pres., Arch C. Gerlach; Sec., Meredith F. Burrill; Treas., George F. Deasy; Councilors, Jan O.M. Broek, Fred B. Kniffen, Alfred H. Meyer, James J. Parsons, Arthur H. Robinson, John F. Lounsbury, Evelyn L. Pruitt.

1962 – Pres., Arch C. Gerlach; Vice-Pres., Arthur H. Robinson; Sec., Meredith F. Burrill; Treas., Leonard S. Wilson; Councilors, Gilbert F. White, James J. Parsons, John F. Lounsbury, Evelyn L. Pruitt, Thomas R. Smith, Herold J. Wiens.

1963 – Pres., Arthur H. Robinson; Vice-Pres., Edward B. Espenshade, Jr.; Sec., Norton S. Ginsburg; Treas., Leonard S. Wilson; Councilors, Arch C. Gerlach, John F. Lounsbury, Evelyn L. Pruitt; Thomas R. Smith, Herold J. Wiens, John R. Borchert, Rhoads Murphey.

1964 – Pres., Edward B. Espenshade, Jr.; Vice-Pres., Meredith F. Burrill; Sec., Norton S. Ginsburg; Treas., Leonard S. Wilson; Councilors, Arthur H. Robinson, Thomas R. Smith, Herold J. Wiens, John R. Borchert, Rhoads Murphey, Donald W. Meinig, Dan Stanislawski; Executive Officer, Arvin W. Hahn.

1965 – Pres., Meredith F. Burrill; Vice-Pres., Walter M. Kollmorgen; Sec., Norton S. Ginsburg; Treas., Alvin A. Munn; Past Pres., Edward B. Espenshade; Councilors, John R. Borchert, Rhoads Murphey, Harold M. Mayer, Wilbur Zelinsky,

Donald W. Meinig, Dan Stanislawski; Executive Officer, Saul B. Cohen.

1966 – Pres., Walter M. Kollmorgen; Vice-Pres., Clyde F. Kohn; Sec., John P. Augelli; Treas., Alvin A. Munn; Past Pres., Meredith F. Burrill; Councilors, J. Ross Mackay, Wilbur Zelinsky, Donald W. Meinig, Edward J. Taaffe, Dan Stanislawski, Harold M. Mayer; Executive Officer, John Fraser Hart. (Prior to 1967, chairpersons of Regional Divisions served as Councilors for one year terms. From 1967 each division elected its councilor to serve a three year term on the AAG Council.)

1967 – Pres., Clyde F. Kohn; Vice-Pres., John R. Borchert; Sec., John P. Augelli; Treas., Alvin A. Munn; Past Pres., Walter M. Kollmorgen; Councilors, Douglas B. Carter, J. Ross Mackay, Edward J. Taaffe, Saul B. Cohen, Wilbur Zelinsky, Harold M. Mayer; Regional Councilors, Walter W. Deshler-MAD; Merle C. Prunty-SE; Edwin H. Hammond-MS; Lawrence M. Sommers-EL; Donn K. Haglund-WL; Joseph Castelli-GP/RM; C. Miller Strack-SW; Howard J. Nelson-PC; George K. Lewis-NE/St.L; Executive Secretary, J. Warren Nystrom.

1968 – Pres., John R. Borchert; Vice-Pres., J. Ross Mackay; Sec., John P. Augelli; Treas., Robert H. Alexander; Past Pres., Clyde F. Kohn; Councilors, Douglas B. Carter, Robert A. Harper, David Lowenthal, Edward J. Taaffe, Saul B. Cohen, Neil E. Salisbury; Regional Councilors, James A. Brammell-MAD; Merle C. Prunty-SE; Edwin H. Hammond-MS; Lawrence M. Sommers-EL; Donn K. Haglund-WL; Leslie Hewes-GP/RM; Lorrin Kennamer-SW; Howard J. Nelson-PC; George K. Lewis-NE/St.L; Executive Secretary, J. Warren Nystrom.

1969 – Pres., J. Ross Mackay; Vice-Pres., Norton S. Ginsburg; Sec., Wesley C. Calef; Treas., Robert D. Hodgson; Past Pres., John R. Borchert; Councilors, Douglas B. Carter, Robert A. Harper, David Lowenthal, Clyde P. Patton, Saul B. Cohen, Harold M. Rose; Regional Councilors, James A. Brammell-MAD; Merle C. Prunty Jr., SE; Edwin H. Hammond-MS; Lawrence M. Sommers-EL; Donn K. Haglund-WL; Leslie Hewes-GP/RM; Lorrin Kennamer-SW; Howard J. Nelson-PC; Emanuel Maier-NE/St. L; Executive Secretary, J. Warren Nystrom.

1970 – Pres., Norton S. Ginsburg; Vice-pres., Edward J. Taaffe; Sec., Wesley C. Calef; Treas., Robert D. Hodgson; Past Pres., J. Ross Mackay; Councilors, Robert A. Harper, David Lowenthal, Clyde P. Patton, Harold M. Rose, Harm J. de Blij, Tom McKnight; Regional Councilors, James A. Brammel-MAD; James R. Anderson-SE; Russell White-MS; Lawrence M. Sommers-EL; Donn K. Haglund-WL; Leslie Hewes-GP/RM; Lorrin Kennamer-SW; Howard J. Nelson-PC; Emanuel Maier-NE/St. L; Executive Director, J. Warren Nystrom.

1971 – Pres., Edward J. Taaffe; Vice Pres., Wilbur Zelinsky; Sec., Wesley C. Calef; Treas., Donald W. Wise; Past Pres., Norton S. Ginsburg; Councilors, Clyde P. Patton, Harold M. Rose, Harm J. de Blij, Tom McKnight, Richard L. Morrill, Philip L. Wagner; Regional Councilors, Gordon C. Hull-MAD; James R. Anderson-SE; Leonard Zobler-MS; Robert B. McNee-EL; Howard G. Roepke-WL; Leslie Hewes-GP/RM; Lorrin Kennamer-SW; Howard J. Nelson-PC; Emanuel Maier-NE/St. L; Executive Director, J. Warren Nystrom.

1972 – Pres., Wilbur Zelinsky; Vice-Pres., Julian Wolpert; Sec., Harm J. de Blij; Treas., Donald W. Wise; Past Pres., Edward J. Taaffe; Councilors, Ian Burton, Tom McKnight, Richard L. Morrill, Philip L. Wagner, Barbara Borowiecki, Marvin W. Mikesell; Regional Councilors, Gordon C. Hull-MAD; James R. Anderson-SE; Leonard Zobler-MS; Robert B. McNee-EL; Howard G. Roepke-WL; Richard E. Lonsdale-GP/RM; Allen D. Hellman-SW; William L. Thomas-PC; William H. Wallace-NE/St.L; Executive Director, J. Warren Nystrom.

1973 – Pres., Julian Wolpert; Vice Pres., James J. Parsons; Sec., Harm J. de Blij; Treas., Donald W. Wise; Past Pres., Wilbur Zelinsky; Councilors, Richard L. Morrill, Philip L. Wagner, Barbara Borowiecki, Marvin W. Mikesell, Elinore M. Barrett, David W. Harvey; Regional Councilors, Gordon C. Hull-MAD; James A. Shear-SE; Leonard Zobler-MS; Robert B. McNee-EL, Howard G. Roepke-WL; Richard E. Lonsdale-GP/RM; Allen D. Hellman-SW; William I. Thomas-PC; William H. Wallace-NE/St.l.; Executive Director, J. Warren Nystrom.

1974 – Pres., James J. Parsons; Vice-Pres., Marvin W. Mikesell; Sec., Harm J. de Blij; Treas., Dorothy A. Nicholson; Past

Pres., Julian Wolpert; Councilors, Barbara Borowiecki, Elinore M. Barrett, David W. Harvey, Yi-Fu Tuan, James E. Vance, Morgan D. Thomas; Regional Councilors, Evelyn L. Pruitt-MAD; James A. Shear-SE; Russell A. White-MS; Donald R. Deskins-EL; Herbert H. Gross-WL; Richard E. Lonsdale-GP/RM; H. Jesse Walker-SW; Norman J. Thrower-PC; William H. Wallace-NE/St.L; Executive Director, J. Warren Nystrom.

1975 – Pres., Marvin W. Mikesell; Vice Pres., Harold M. Rose; Sec., John S. Adams; Treas., Dorothy A. Nicholson; Past Pres., James J. Parsons; Councilors, Elinore M. Barrett, David W. Harvey, Yi-Fu Tuan, James E. Vance, Annette Buttimer, David Ward; Regional Councilors, Evelyn L. Pruitt-MAD; James A. Shear-SE; Russell A. White-MS; Donald R. Deskins-EL; Herbert H. Gross-WL; Nicholas F. Helburn-GP/ RM; H. Jesse Walker-SW; Norman J. Thrower-PC; Edward J. Miles-NE/St.L; Executive Director, J. Warren Nystrom.

1976 – Pres., Harold M. Rose; Vice-Pres., Melvin G. Marcus; Sec., John S. Adams; Treas., Dorothy A. Nicholson; Past Pres., Marvin W. Mikesell; Councilors, Yi-Fu Tuan, James E. Vance, Annette Buttimer, David Ward, Robert B. McNee, Risa I. Palm; Regional Councilors, Evelyn L. Pruitt-MAD; David G. Basile-SE; Kennard W. Rumage-MS; Kenneth E. Corey-EL; John Fraser Hart-WL; Nicholas F. Helburn-GP/ RM; Phillip Bacon-SW; Norman J. Thrower-PC; Edward J. Miles-NE/St.L; Executive Director, J. Warren Nystrom.

1977 – Pres., Melvin G. Marcus; Vice-Pres., Brian J.L. Berry; Sec., John S. Adams; Treas., Thomas J. Wilbanks; Past Pres., Harold M. Rose; Councilors, Annette Buttimer, David Ward, Robert B. McNee, Risa I. Palm, James R. Anderson, Richard E. Lonsdale; Regional Councilors, Raymond S. Honda-MAD; David G. Basile-SE, Kennard W. Rummage-MS; Kenneth E. Corey-EL; John Fraser Hart-WL; Nicholas F. Helburn-GP/RM; Phillip Bacon-SW; Everett G. Smith, Jr.-PC; Edward J. Miles-NE/St.L; Executive Director, J. Warren Nystrom.

1978 – Pres., Brian J.L. Berry; Vice Pres., John Fraser Hart; Sec., Richard L. Morrill; Treas., Thomas J. Wilbanks; Past Pres., Melvin G. Marcus; Councilors, Robert B. McNee, Risa I. Palm, James R. Anderson, Richard E. Lonsdale, Lawrence A.

Brown, Janice Monk; Regional Councilors, Raymond S. Honda-MAD; James O. Wheeler-SE; Kennard W. Rumage-MS; Kenneth E. Corey-EL; Alice M.T. Rechlin-WL; Jacquelyn L. Beyer-GP/RM; Phillip Bacon-SW; Everett G. Smith Jr.-PC; Terence Burke-NE/St.L; Executive Director, J. Warren Nystrom.

APPENDIX C

PRESIDENTS OF THE ASSOCIATION OF AMERICAN GEOGRAPHERS: ANNUAL ADDRESS

1904 – Philadelphia, Pennsylvania. This initial meeting of the Association included a program of some twenty papers. It was presided over by William M. Davis who read a short prologue, the subject of which was "The Opportunity for the Association of American Geographers," published in the *Bulletin of the American Geographical Society,* Vol. 37, pp. 84-86, 1905.

1905 – New York City. Address by William M. Davis: "An Inductive Study of the Content of Geography," *Bulletin of the American Geographical Society,* Vol. 38, pp. 67-84, 1906.

1906 – New York City. Address by Cyrus C. Adams: "Some Phases of Future Geographical Work in America," *Bulletin of the American Geographical Society,* Vol. 39, pp. 1-12, 1907.

1907 – Chicago, Illinois. Address not given because of the death of President Angelo Heilprin late in his term of office.

1908 – Baltimore, Maryland. Address by Grove K. Gilbert: "Earthquake Forecasts," *Science,* Vol. 29, pp. 121-138, 1909.

1909 – Cambridge-Boston, Massachusetts. Address by William M. Davis: "The Italian Riviera Levante." Apparently not published.

1910 – Pittsburgh, Pennsylvania. Address by Henry C. Cowles: "The Causes of Vegetational Cycles," *Annals,* Vol. 1, pp. 3-20, 1911.

1911 – Washington, D.C. Address by Ralph S. Tarr: "The Glaciers of Alaska," *Annals,* Vol. 2, pp. 3-24, 1912.

1912 – New Haven, Connecticut. Address apparently not given. Rollin D Salisbury was absent from the meetings.

1913 – Princeton, New Jersey. Address by Henry G. Bryant: "Government Agencies and Geography in the United States." Apparently not published.

1914 – Chicago, Illinois. Address by Albert P. Brigham: "Problems of Geographic Influence," *Annals,* Vol. 5, pp. 3-25, 1915.

1915 – Washington, D.C. Address by Richard E. Dodge: "Some Problems in Geographic Education, with Special Reference to Secondary Schools," *Annals,* Vol. 6, pp. 3-18, 1916.

1916 – New York City. Address by Mark Jefferson: "Geographic Provinces of the United States," *Annals,* Vol. 7, pp. 3-15, 1917.

1917 – No meeting. Scheduled to be held in Chicago, Illinois, this meeting was cancelled because of wartime restrictions on travel. Address by Robert DeC. Ward: "Meteorology and War-Flying, Some Practical Suggestions," *Annals,* Vol. 8, pp. 3-33, 1918.

1918 – Baltimore, Maryland. Address by Nevin M. Fenneman: "The Circumference of Geography," *Annals,* Vol. 9, pp. 3-11, 1919.

1919 – St. Louis, Missouri. Address by Charles R. Dryer: "Genetic Geography: The Development of the Geographic Sense and Concept," *Annals,* Vol. 10, pp. 3-16, 1920.

1920 – Chicago, Illinois. Herbert E. Gregory: "Geographic Basis of the Political Problems of the Pacific." Apparently read by title only.

1921 – Washington, D.C. Address by Ellen C. Semple: "The Influence of Geographic Conditions upon Current Mediterranean Stock-Raising," *Annals,* Vol. 12, pp. 3-38, 1922.

1922 – Ann Arbor, Michigan. Address by Harlan H. Barrows: "Geography as Human Ecology," *Annals,* Vol. 13, pp. 1-14, 1923.

1923 – Cincinnati, Ohio. Address by Ellsworth Huntington: "Geography and Natural Selection: A Preliminary Study of the Origin and Development of Racial Character," *Annals,* Vol. 14, pp. 1-16, 1924.

1924 – Washington, D.C. Address by Curtis F. Marbut: "The Rise, Decline, and Revival of Malthusianism in Relation to Geography and the Character of Soils," *Annals,* Vol. 15, pp. 1-29, 1925.

1925 – Madison, Wisconsin. Address by Ray H. Whitbeck: "Adjustments to Environment in South America: An Interply (Interplay) of Influences," *Annals,* Vol. 16, pp. 1-11, 1926.

1926 – Philadelphia, Pennsylvania. Address by J. Paul Goode: "The Map as a Record of Progress in Geography," *Annals,* Vol. 17, pp. 1-14, 1927.

1927 – Nashville, Tennessee. Address by Marius R. Campbell: "Geographic Terminology," *Annals,* Vol. 18, pp. 25-40, 1928.

1928 – New York City. Address by Douglas W. Johnson: "The Geographic Prospect," *Annals,* Vol. 19, pp. 167-231, 1929.

1929 – Columbus, Ohio. Address by Lawrence Martin: "The Michigan-Wisconsin Boundary Case in the Supreme Court of the United States, 1923-26," *Annals,* Vol. 20, pp. 105-163, 1930.

1930 – Worcester, Massachusetts. Address by Almon E. Parkins: "The Antebellum South: A Geographer's Interpretation," *Annals,* Vol. 21, pp. 1-33, 1931.

1931 – Ypsilanti, Michigan. Address by Isaiah Bowman: "Planning in Pioneer Settlement," *Annals,* Vol. 22, pp. 93-107, 1932.

1932 – Washington, D.C. Address by Oliver E. Baker: "Rural-Urban Migration and the National Welfare," *Annals,* Vol. 23, pp. 59-126, 1933.

1933 – Evanston, Illinois. Address by François E. Matthes: "Our Greatest Mountain Range, the Sierra Nevada of California." Apparently not published.

1934 – Philadelphia, Pennsylvania. Address by Wallace W. Atwood: "The Increasing Significance of Geographic Conditions in the Growth of Nation States," *Annals,* Vol. 25, pp. 1-16, 1935.

1935 – St. Louis, Missouri. Address by Charles C. Colby: "Changing Currents of Geographic Thought in America," *Annals,* Vol. 26, pp. 1-37, 1936.

1936 – Syracuse, New York. Address by William H. Hobbs: "The Progress of Discovery and Exploration within the Arctic Region," *Annals,* Vol. 27, pp. 1-22, 1937.

1937 – Ann Arbor, Michigan. Address by W.L.G. Joerg programmed as "Generalization and Synthesis in Geography," was not given because of the absence (due to illness) of the President. In its place Robert B. Hall gave an address, "The Expansion of the Japanese Empire." The Presidential Address has not been published.

1938 – Cambridge, Massachusetts. Address by Vernor C. Finch: "Geographical Science and Social Philosophy," *Annals,* Vol. 29, pp. 1-28, 1939.

1939 – Chicago, Illinois. Address by Claude H. Birdseye: "Stereoscopic Phototopographic Mapping," *Annals,* Vol. 31, pp. 1-24, 1940.

1940 – Baton Rouge, Louisiana. Address by Carl O. Sauer: "Foreword to Historical Geography," *Annals,* Vol. 31, pp. 1-24, 1941.

1941 – New York City. Address by Griffith Taylor: "Environment, Village and City: A Genetic Approach to Urban Geography, with Some Reference to Possibilism," *Annals,* Vol. 32, pp. 1-67, 1942.

1942 – No meeting. The scheduled meetings were postponed in the interest of curtailing railroad travel in wartime. President J. Russell Smith's address was given the following year at the September 1943 Annual Meeting in Washington, D.C.: "Grassland and Farmland as Factors in the Cyclical Development of Eurasian History," *Annals,* Vol. 33, pp. 135-161, 1943.

1943 – Washington, D.C. Address by Hugh H. Bennett: "Adjustment of Agriculture to Its Environment," *Annals,* Vol. 33, pp. 163-195, 1943.

1944 – The Annual Meetings were postponed in the interest of curtailing railroad travel in wartime. However, President Derwent Whittlesey prepared a paper: "The Horizon of Geography," which was published in the *Annals,* Vol. 35, pp. 1-36, 1945.

1945 – Knoxville, Tennessee. Address by Robert S. Platt: "Problems of Our Times," *Annals,* Vol. 36, pp. 1-43, 1946.

1946 – Columbus, Ohio. Address by John K. Wright: "Terrae Incognitae: The Place of Imagination in Geography," *Annals,* Vol. 37, pp. 1-15, 1947.

1947 – Charlottesville, Virginia. Address by Charles F. Brooks: "The Climatic Record: Its Content, Limitations, and Geographic Value," *Annals,* Vol. 38, pp. 153-168, 1948.

1948 – Madison, Wisconsin. Address by Richard J. Russell: "Geographical Geomorphology," *Annals,* Vol. 38, pp. 1-11, 1949.

1949 – No meeting was held in 1949, the Council having authorized the change from December 1949 to April 1950.

1950 – Worcester, Massachusetts. Address by Richard Hartshorne: "The Functional Approach in Political Geography," *Annals,* Vol. 40, pp. 95-130, 1950.

1951 – Chicago, Illinois. Address by G. Donald Hudson: "Professional Training of the Membership of the Association of American Geographers," *Annals,* Vol. 41, pp. 97-115, 1951.

1952 – Washington, D.C. Address by Preston E. James: "Toward a Further Understanding of the Regional Concept," *Annals,* Vol. 42, pp. 195-222, 1952.

1953 – Cleveland, Ohio. Address by Glenn T. Trewartha: "A Case for Population Geography," *Annals,* Vol. 43, pp. 71-97, 1953.

1954 – Philadelphia, Pennsylvania. Address by J. Russell Whitaker: "The Way Lies Open," *Annals,* Vol. 44, pp. 231-244, 1954.

1955 – Memphis, Tennessee. In accordance with the changes in the By-laws the President of the Association ceased to be responsible for prHparing and giving an address; rather, the Honorary President gave the address at the Annual Dinner Meeting. On April 13, 1955, (Derwent Whittlesey, the first Honorary President, gave his address: "Southern Rhodesia – An African Compage," *Annals,* Vol. 46, pp. 1-97, 1956.

1956 – Montreal, Canada. Address by the Honorary President Carl O. Sauer on "The Education of a Geographer," *Annals,* Vol. 46, pp. 287-299, 1956.

1957 – Cincinnati, Ohio. Address by the Honorary President George B. Cressey: "Water in the Desert," *Annals,* Vol. 47, pp. 105-124, 1957.

1958 – Santa Monica, California. Address by the Honorary President John B. Leighly on "John Muir's Image of the West," *Annals,* Vol. 48, pp. 309-318, 1958.

1959 – Pittsburgh, Pennsylvania. Address by the Honorary President Stephen B. Jones on "Boundary Concepts in the Setting of Place and Time," *Annals,* Vol. 49, pp. 241-255, 1959.

1960 – Dallas, Texas. Address by the Honorary President John E. Orchard on "Industrialization of Japan, Mainland China, and India in Its World Significance," *Annals,* Vol. 50, pp. 193-215, 1960.

1961 – East Lansing, Michigan. Address by the Honorary President C. Warren Thornthwaite: "The Task Ahead," *Annals,* Vol. 51, (1961), pp. 345-356.

1962 – Miami Beach, Florida. Address by the Honorary President Andrew H. Clark: "Praemia Geographicae: The Incidental Rewards of a Professional Career," *Annals,* Vol. 52, (1962), pp. 229-241.

1963 – Denver, Colorado. Address by the Honorary President Edward A. Ackerman: "Where is a Research Frontier?" *Annals,* Vol. 53, (1963), pp. 429-440.

1964 – Syracuse, New York. Address by the Honorary President F. Kenneth Hare: "New Light From Labrador – Ungava," *Annals,* Vol. 54, (1964), pp. 459-476.

1965 – Columbus, Ohio. Address by the Honorary President Fred B. Kniffen: "Folk Housing, Key to Diffusion," *Annals,* Vol. 55 (1965), pp. 549-577.

1966 – Toronto, Canada. Address by the Honorary President Preston E. James: "On the Origin and Persistence of Error in Geography," *Annals,* Vol. 57 (1967), pp. 1-24.

1967 – St. Louis, Missouri. Address by the Past President Meredith F. Burrill: "The Language of Geography," *Annals,* Vol. 58, (1968), pp. 1-11.

1968 – Washington, D.C. Address by the Past President, Walter M. Kollmorgen: "The Woodman's Assault on the Domain of the Cattleman," *Annals,* Vol. 59 (1969), pp. 215-239.

1969 – Ann Arbor, Michigan. Address by the Past President, Clyde F. Kohn: "The 1960s: A Decade of Progress in Geographical Research and Instruction," *Annals,* Vol. 60, (1970), pp. 211-219.

1970 – San Francisco, California. Address by the Past President, John R. Borchert: "The 'Dust Bowl' in the 1970s," *Annals,* Vol. 61, (1971), pp. 1-22.

1971 – Boston, Massachusetts. Address by the Past President, J. Ross Mackay: "The World of Underground Ice," *Annals,* Vol. 62 (1972), pp. 1-22.

1972 – Kansas City, Missouri. Address by the Past President, Norton S. Ginsburg: "Colonialism to National Development – Geographic Perspective on Patterns and Policies," *Annals,* Vol. 63 (1973), pp. 1-21.

1973 – Atlanta, Georgia. Address by the Past President, Edward J. Taaffe: "The Spatial View in Context," *Annals.* Vol. 64 (1974), pp. 1-16.

1974 – Seattle, Washington. Address by the Past President, Wilbur Zelinsky: "The Demigod's Dilemma," *Annals,* Vol. 65 (1975), pp. 123-143.

1975 – Milwaukee, Wisconsin. Address by the Past President, Julian Wolpert: "Opening Closed Spaces," *Annals,* Vol. 66, (1976), pp. 1-13.

1976 – New York, New York. Address by the Past President, James J. Parsons: "Geography as Exploration and Discovery," *Annals,* Vol. 67 (1977), pp. 1-16.

1977 – Salt Lake City, Utah. Address by the Past President, Marvin W. Mikesell, "Tradition and Innovation in Cultural Geography," *Annals,* Vol. 68, (1978), pp. 1-16.

1978 – New Orleans, Louisiana. Address by the Past President, Harold M. Rose: "The Geography of Despair: An Assessment of Rising Levels of Lethal Violence," (in press).

APPENDIX D

OFFICERS OF THE AMERICAN SOCIETY
FOR PROFESSIONAL GEOGRAPHERS

1944 – President, F. Webster McBryde; Secretary, E. Willard Miller; Treasurer, George F. Deasy.

1945 – President, William Van Royen; Vice-President, John K. Rose; Secretary, E. Willard Miller; Treasurer, George F. Deasy.

1946 – President, John K. Rose; Vice-President, W. Elmer Ekblaw; Secretary, E. Willard Miller; Treasurer, George F. Deasy.

1947 – President, Otis P. Starkey; Vice-President, Sidman P. Poole; Secretary, E. Willard Miller; Treasurer, George F. Deasy.

1948 – President, E. Willard Miller; Vice President, Louis O. Quam; Secretary, Walter W. Ristow; Treasurer, George F. Deasy.

APPENDIX E

THE HISTORY OF THE ASSOCIATION OF AMERICAN GEOGRAPHERS: REFERENCES TO SOME PUBLISHED SOURCES.

BRIGHAM, Albert P.: "The Association of American Geographers, 1903-1923," *Annals*, Vol. 14, (1924), pp. 109-116.

COLBY, Charles C.: "Twenty-Five Years of the Association of American Geographers: A Secretarial Review," *Annals*, Vol. 19, (1929), pp. 59-61.

DAVIS, William M.: "Geography in the United States," *Science*, New Series, Vol. 19, (1904), pp. 120-132, 178-186.

DAVIS, William M.: "The Opportunity for the Association of American Geographers," *Bulletin of the American Geographical Society*, Vol. 37, (1905), pp. 84-86.

DAVIS, William M.: "The Progress of Geography in the United States," *Annals*, Vol. 14, (1924), pp. 159-215.

DRYER, Charles R.: "A Century of Geographic Education in the United States," *Annals*, Vol. 14, (1924), pp. 117-149.

FRIIS, Herman R.: "The Distribution of Members of the Association of American Geographers and of the Middle Atlantic Division, December 1953: A Cartographic Presentation," *The Professional Geographer*, New Series, Vol. 6, maps and tables, (1954) pp. 6-14.

HARTSHORNE, Richard: "The President Speaks," *The Professional Geographer*, New Series, Vol. 1, (1949), pp. 17-22.

HUDSON, G. Donald: "Professional Training of the Membership of the Association of American Geographers," *Annals*, Vol. 41, (1951), pp. 97-115.

JONES, Clarence F.: "Status and Trends of Geography in the United States, 1952-1957; A Report Prepared for the Commission on Geography, Pan American Institute of Geography and History," *The Professional Geographer,* New Series, Vol. 11 (1), Part 2, pp. 1-145, (1959). See especially "Association of American Geographers," pp. 122-123.

PATTISON, William D.: "The Star of the AAG," *The Professional Geographer,* New Series, Vol. 12, (1960), pp. 18-19.

PLATT, Robert S.: "Finances of the Association of American Geographers," *Annals,* Vol. 22, (1932), pp. 91-92.

SMITH, J. Russell: "American Geography: 1900-1904," *The Professional Geographers,* New Series, Vol. 4, (1952); pp. 4-7.

WRIGHT, John K.: *Geography in the Making: The American Geographical Society,* (New York, 1952), 457 pp. See pp. 143-144, 166-168, 223, 270.

WRIGHT, John K.: "AAG Programs and Program-Making 1904-1954," *The Professional Geographer,* New Series, Vol. 6, pp. 6-11, (1954).

Anonymous: "The Association of American Geographers," *Bulletin of the American Geographical Society,* Vol. 37, (1905), p. 42.

Anonymous: "War Services of Members of the AAG," *Annals,* Vol. 9, (1919), pp. 49-70.

APPENDIX F

THE HISTORY OF THE AMERICAN SOCIETY FOR PROFESSIONAL GEOGRAPHERS: REFERENCES TO SOME PUBLISHED SOURCES.

DEASY, George F.: "Training, Professional Work, and Military Experience of Geographers, 1942-1947," *Bulletin of the American Society for Professional Geographers,* Vol. 6, (December, 1947), pp. 1-14.

DEASY, George F.: "War-Time Changes in Occupation of Geographers," *Bulletin of the American Society for Professional Geographers,* Vol. 7, (April, 1948), pp. 33-41.

MILLER, E. Willard: "A Short History of the American Society of Professional Geographers," *The Professional Geographer,* New Series, Vol. 2, (1950), pp. 29-40.

VAN ROYEN, William: "The Functions of Our Society," *Bulletin of the American Society for Professional Geographers,* Vol. 1-2, (January - March, 1945), pp. 1-3.

Anonymous: "Description of the Professional Occupation of Geography," *Bulletin of the American Society for Professional Geographers,* Vol. 2 (1), (April, 1944), pp. 1-2.

Anonymous: "Status of Professional Geographers in the United States – December 1943," *Bulletin of the American Society for Professional Geographers,* Vol. 2 (2), May, 1944), pp. 1-2.

APPENDIX G

MAPS OF THE MEMBERSHIP OF THE ASSOCIATION OF AMERICAN GEOGRAPHERS: 1904, 1919, 1949 AND 1978

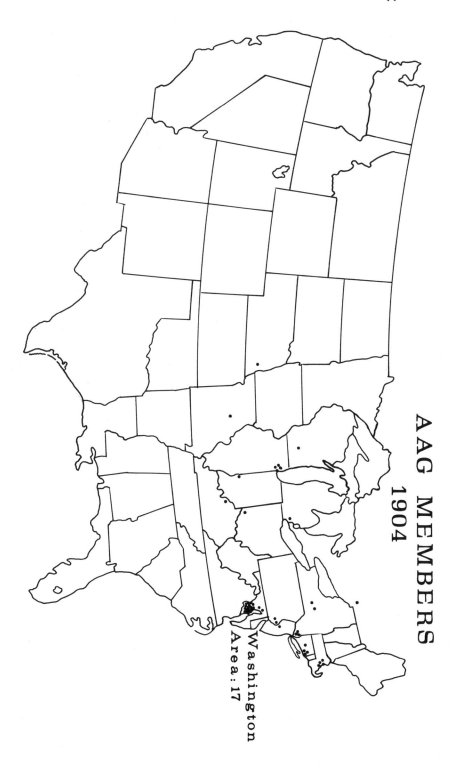

AAG MEMBERS
1904

Washington
Area: 17

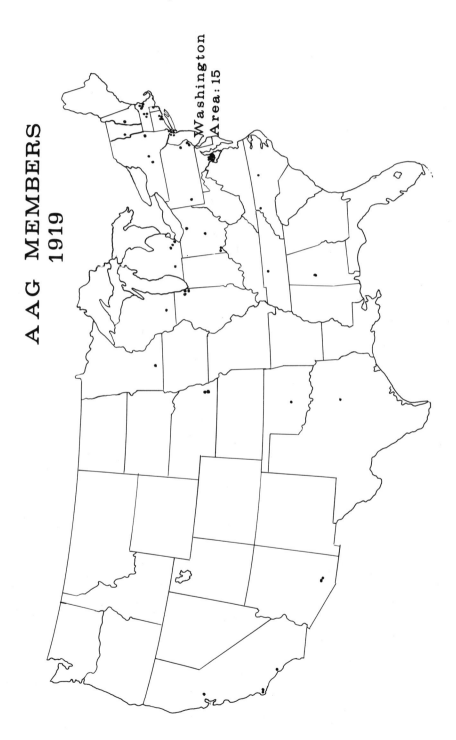

AAG MEMBERS 1919

Washington Area: 15

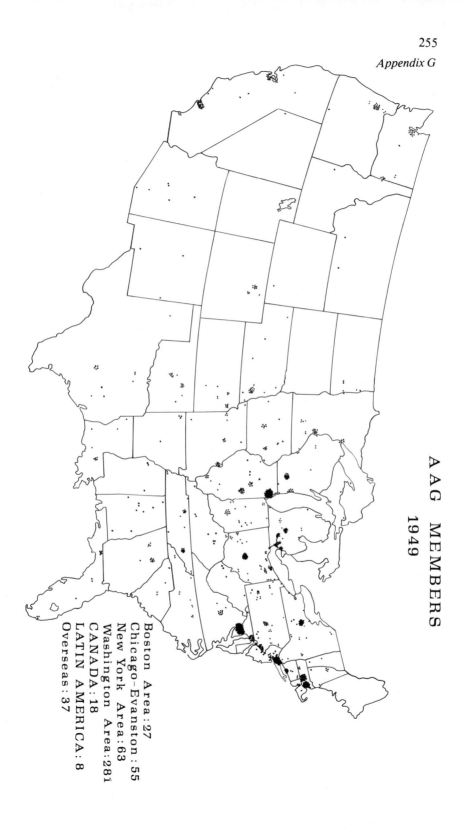

A A G M E M B E R S
1949

Boston Area:27
Chicago-Evanston:55
New York Area:63
Washington Area:281
CANADA:18
LATIN AMERICA: 8
Overseas:37

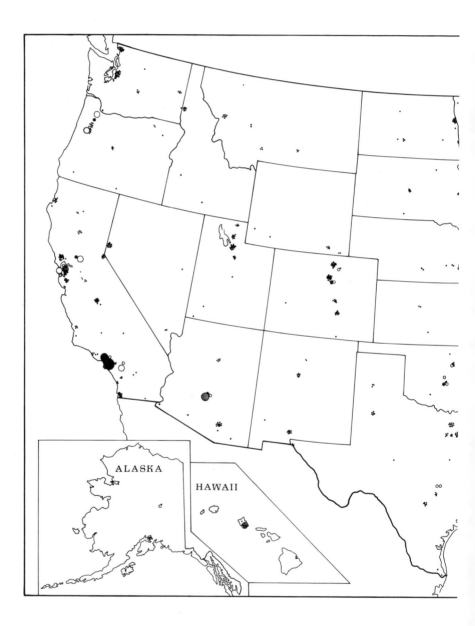

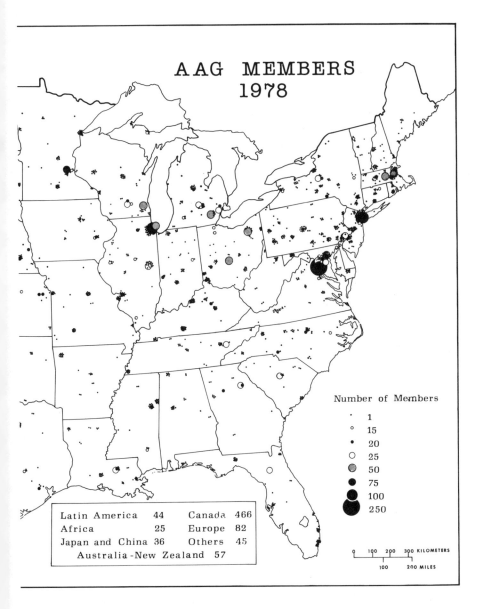

AAG MEMBERS
1978

Number of Members

.   1
○   15
•   20
○   25
⬤   50
●   75
●   100
●   250

Latin America	44	Canada	466
Africa	25	Europe	82
Japan and China	36	Others	45
Australia-New Zealand	57		

0   100   200   300 KILOMETERS

100   200 MILES

APPENDIX H

MAP OF GRADUATE DEPARTMENTS OF GEOGRAPHY IN THE UNITED STATES, 1979.

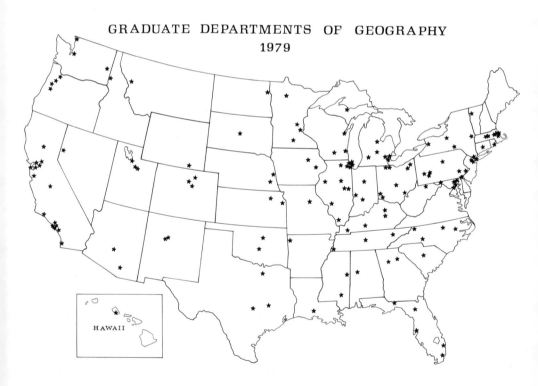

GRADUATE DEPARTMENTS OF GEOGRAPHY
1979

HAWAII

APPENDIX I

ASSOCIATION OF AMERICAN GEOGRAPHERS: HONORS AWARDED

The AAG Honors Awards are "in special recognition of outstanding contributions toward the advancement or welfare of the profession" (AAG Bylaws, sec. 2).

The first recipient of a Distinguished Achievement Award was Gladys M. Wrigley in 1951. Thereafter this award was discontinued and Honors have been granted as indicated below:

Year	Outstanding Achievement	Meritorious Contributions
1952	None	Ackerman, Edward A.
		Black, Lloyd D.
		Jenks, George F.
		LeGear, Clara E.
1953	Thornthwaite, C. Warren	Atwood Jr., Wallace W.
		Kollmorgen, Walter M.
		May, Jacques M.
		Robinson, Arthur H.
1954	Shantz, Homer LeRoy	Marschner, Francis J.
		Murphy, Raymond E.
		Watson, James Wreford
1955	White, Gilbert F.	Kendall, Henry M.
		Raisz, Erwin
		Weaver, John C.
1956	Wright, John K.	Roterus, Victor
		Hall, Robert Burnett
1957	None	None
1958	Pico, Rafael	Carter, George F.
		Johnson, Hildegard B.
		Ullman, Edward L.
1959	Visher, Stephen S.	Applebaum, William
		Barnes, Carleton P.
		Batschelet, Clarence E.
		Ginsburg, Norton S.

1960	Hartshorne, Richard	Garrison, William L.
	Russell, Richard J.	Lowenthal, David
1961	Hitchcock, Charles B.	Espenshade Jr., Edward B.
		Hare, F. Kenneth
		Thomas, William L.
1962	None	Gottmann, Jean
		Kimble, George H. T.
		Kollmorgen, Walter M.
1963	Grosvenor, Gilbert H.	Jackson, John B.
		Prunty, Merle C.
		Stanislawski, Dan
1964	Christaller, Walter	Fairchild, Wilma B.
		Mackay, J. Ross
		West, Robert C.
1965	Friis, Herman R.	Guthe, Otto E.
	McCarty, Harold H.	Hewes, Leslie
		Meinig, Donald W.
		Stone, Kirk H.
1966	Siple, Paul A.	Robinson, J. Lewis
		Van Royen, William
		Zelinsky, Wilbur
1967	None	Harrison, Richard E.
		Schwendeman Sr., Joseph R.
		Wilson, Leonard S.
1968	Hagerstrand, Torsten	Berry, Brian J.L.
	Spencer, Joseph E.	Butzer, Karl W.
		Glacken, Clarence J.
		Hammond, Edwin H.
1969	Gerlach, Arch C.	Hart, John Fraser
		Shabad, Theodore
		Curry, Leslie
1970	None	None
1971	Ajo, Reino	Dacey, Michael F.
		Morrill, Richard L.
		Tobler, Waldo R.
1972	Dickinson, Robert E.	Aschmann, Homer
		Pruitt, Evelyn L.
		Wolman, M. Gordon
1973	West, Robert C.	Haggett, Peter
		Sherman, John C.
		Tuan, Yi-Fu
1974	White, Gilbert F.	Bertrand, Kenneth J.
	(Special Award)	Burrill, Meredith F.
		Simonett, David S.
		Wheatley, Paul
		Sauer, Carl O.
1975	Trewartha, Glenn T.	DeVorsey Jr., Louis
		Dury, George H.
		Gould, Peter R.

In 1975, as a result of mail ballot to the membership, the Bylaws were changed, eliminating the categories of "Outstanding Achievement" and "Meritorious Contribution."

1976 Honors Awards	Borchert, John R. Harris, Chauncy D. King, Leslie J.
1977	Darby, H. Clifford Leighly, John B. Lewis, Peirce F. Walker, H. Jesse
1978	Kniffen, Fred B. Küchler, A. William Pred, Allan

Index

A

Abbe, Cleveland, 205
Abbe, Cleveland, Jr., 36
Abler, R. F., 170
Ackerman, E. A., 84, 98, 132
Adams, Charles C., 33, 36
Adams, Cyrus C., 29, 38, 41,
  55, 58; original member,
  AAG, 36; Councilor, AAG,
  40; papers read, 41; presiden-
  tial address, 53
Adams, J. S., 170
Adkinson, B. W., 109
Agassiz, Louis, 26
Agassiz Association, 27
Alaska Geographical Society, 188
Alexander, L. M., 131
Allen, J. A., 38
American Academy of Arts and
  Sciences, 3
American Alpine Club, 27
American Anthropological
  Association, 3
American Association for the
  Advancement of Science, 31–
  32, 34, 136, 145–146, 178;
  AAG representatives, 214;
  founding of, 3, 26; joint
  meetings with AAG, 40, 49
American Astronomical
  Society, 98

American Chemical Society, 98
American Congress on Surveying
  and Mapping, 133
American Council of Learned
  Societies, 95, 110, 133, 134,
  136, 175, 178,
American Council on Education, 134; AAG Central Office
  in, 122
American Economic Association, 3, 98
American Folklore Society, 3
American Geographical and
  Statistical Society, 2, 27
American Geographical Society,
  27, 35, 36, 66, 71, 83, 93, 94,
  96, 107, 108, 175, 192, 193,
  199; *Bulletin,* 27, 31, 58, 61,
  190; centennial of, 131, 140;
  *Current Geographical Publi-
  cations,* 177; *Focus,* 177;
  founding of, 27; *Geographical
  Review,* 27, 84, 177; joint
  meetings with AAG, 59, 61,
  62, 72, 190; membership of,
  187;participation in National
  Atlas, 134–135; relations with
  AAG, 58–62, 175, 177; relations with NGS, 29; support
  of *Annals,* AAG, 58–61, 62;
  Transcontinental Excursion,

original member, AAG, 34,
37; papers read, 35
Ganong, W. F., 205
Garland, J. H., 139
Garrison, W. L., 113, 136, 145,
147
Genthe, M. K., 31; original
member, AAG, 37, 38; papers
read, 50; resignation from
AAG, 47
Geographers, in business, 141,
142, 172; employment of, 185;
in government, 141, 142,
172; in World War I, 70,
192; in World War II, 89, 90
Geographic Society of Chicago,
27, 29, 35, 188
Geographical Club of Phila-
delphia, 27
Geographical Journal, 2, 190
Geographical Review. See
American Geographical Society
Geographical societies; in
Europe, 2; in the United
States, 1903, 26. *See also*
individual entries
Geographical Society of
Baltimore, 27, 188
Geographical Society of Cal-
ifornia, 27, 187
Geographical Society of Phila-
delphia, 27, 29, 34,35,187–188
Geographical Society of the
Pacific, 27
Geographische Zeitschrift, 2
Geography: in business, 77, 129;
careers in, 141, 153; cultural,
78; development as discipline,
6–9; doctorates awarded in,
44; early advanced training in,
11; economic, 11, 43, 50, 77,
78; in education, 143–147; in

Europe, 186; field conferences, 64, 72, 75, 193; in Germany, 6–7, 18, 45; in government, 129; growth of, in fifties and sixties, 149; methodology in, 114; nature of, 14,
25, 45–46, 50, 51, 71–72, 79,
113–114; political, 78; population, 78; regional, 77; as scholarly field, 13–14; in secondary
schools, 14, 15, 55, 56, 129,
144–147, 186; as a social science, 16, 71, 80; status of,
141, 142, 149; in the United
States, 186; university students in, 1910, 21; urban, 78;
women in, 184
Geography, 2
Geological Society of America,
32, 49
Geological Survey of California, 5
Geppert, Otto, 191
Gerasimov, I. P., 139
Gerlach, A. C., 108, 109, 122,
135, 141, 145, 157; president,
AAG, 123, 150
Gerland, G. K. C., 6
Gibson, J. S., 216
Gilbert, G. K., 5, 7, 31, 35;
original member, AAG, 37;
papers read, 35, 41; president,
AAG, 47, 52; vice- president,
AAG, 39, 41
Gilman, D. C., 7, 11, 17, 20
Ginsburg, N.S., 130, 138
Glenn, J. R., 182
Goldthwait, J. W., 188
Goode, J. P., 38, 54, 55, 58, 65,
66; at Chicago, 19–20, 24, 61;
death of, 83; dissertation, 20,
22; original member, AAG,